Elasticity of Materials

Edited by Gülşen Akın Evingür
and Önder Pekcan

Published in London, United Kingdom

IntechOpen

Supporting open minds since 2005

Elasticity of Materials
http://dx.doi.org/10.5772/intechopen.95633
Edited by Gülşen Akın Evingür and Önder Pekcan

Contributors
Indrani Sen, S. Sujith Kumar, Bryer C. Casey Sousa, Danielle L. Cote, Jennifer Hay, Seiki Augustine Chiba, Mikio Waki, Shijie Zhu, Tonghuan Qu, Kazuhiro Ohyama, Eusebio Jiménez López, Mario Acosta Flores, Marta Lilia Eraña Díaz, Jeremiah Rushchitsky, Sanjay Pal, Kinsuk Naskar, Mithun Das, Kirill F. Komkov, Ramratan Guru, Rajeev Kumar Varshney, Rohit Kumar, José Moreira de Sousa, Yohichi Kohzuki

First published in London, United Kingdom, 2023 by IntechOpen
IntechOpen is the global imprint of INTECHOPEN LIMITED, registered in England and Wales, registration number: 11086078, 5 Princes Gate Court, London, SW7 2QJ, United Kingdom
Printed in Croatia

British Library Cataloguing-in-Publication Data
A catalogue record for this book is available from the British Library

Additional hard and PDF copies can be obtained from orders@intechopen.com

Elasticity of Materials
Edited by Gülşen Akın Evingür and Önder Pekcan
p. cm.
Print ISBN 978-1-83969-960-3
Online ISBN 978-1-83969-961-0
eBook (PDF) ISBN 978-1-83969-962-7

Meet the editors

Dr. Gülşen Akın Evingür is an associate professor at Pîrî Reis University, Istanbul, Turkey. She completed her BSc at Yildiz Technical University, Istanbul, Turkey, in 1996, and her Ph.D. at Istanbul Technical University in 2011. Her research interests are composites and their optical, electrical, and mechanical properties. She has authored more than 55 journal articles and 100 conference proceedings in national and international journals. She has edited four books, authored six book chapters, and has participated in both national and international projects..

Prof. Önder Pekcan received his MSc in physics from the University of Chicago, USA in 1971, and his Ph.D. in physics from the University of Wyoming, USA in 1974. He was a visiting scientist at the Abdus Salam International Centre for Theoretical Physics (ICTP), in Italy, and the Technical University of Gdansk, Poland. He was also a visiting professor in the Department of Chemistry, at the University of Toronto, Canada. Dr. Pekcan was appointed a full professor in the Department of Physics, at Istanbul Technical University, where he worked until 2005. Currently, he is a professor in the Department of Bioinformatics and Genetics, at Kadir Has University, Turkey. He has been a member of the Science Academy since 2012. He has more than 380 journal articles, 28 book chapters, and 10 projects to his credit.

Contents

Preface

Elasticity is the ability of a material body to return to its original shape and size after the removal of a deforming force. The elasticity of materials can be predicted by computational simulations and/or measured in laboratory experiments.

The first section of this book, covering simulation and modeling, contains four chapters. In Chapter 1, Sanjay Pal et al. discuss the elasticity of rubber. Chapter 2, by José Moreira de Sousa, considers failures of nanostructures and fully atomistic molecular dynamics simulations. Kirill Komkov's Chapter 3 describes elements of the nonlinear theory of elasticity based on tensor nonlinear equations. In Chapter 4, Eusebio Jiménez Lopéz et al. describes laminate composite material models.

The second section, on characterization, comprises seven chapters. In Chapter 5, Yohichi Kohzuki highlights the temperature dependence of stress within additive single crystals. Jeremiah Rushcitsky (Chapter 6), illustrates the elasticity of auxetic materials. Chapter 7, by Seiki Chiba et al., covers the improvement of elastomer elongation and output for dielectric elastomers. In Chapter 8, Ramratan Guru et al. describes the functional application of compression and recovery in sportswear fabrics. Chapter 9, by IIndrani Sen and S. Sujith Kumar, discusses the characteristic stress-strain behavior of materials caused by nanoindentation. In Chapters 10 and 11, Bryer Sousa et al. discuss the origins of the Oliver–Pharr instrumented strength microprobe method and continued advancements in nanoindentation.

We would like to take this opportunity to thank all the researchers who have made direct contributions to the writing of this book. We also thank all the editorial staff at IntechOpen, in particular Nera Butigan, Author Service Manager, for her effective editing and support during different stages of the production of this book.

Gülşen Akın Evingür
Faculty of Engineering,
Pîrî Reis University,
Tuzla, İstanbul, Turkey

Önder Pekcan
Faculty of Engineering and Natural Sciences,
Kadir Has University,
Cibali, İstanbul, Turkey

Section 1

Simulation and Modeling

Chapter 1

Origin of Rubber Elasticity

Sanjay Pal, Mithun Das and Kinsuk Naskar

Abstract

Under suitable conditions, virtually all rubbery materials exhibit the ability to sustain deformations followed by complete recovery upon removal of the stress. This phenomenon holds significance beyond the narrow confines of the term "rubber elasticity" This elasticity theory is also of great importance in the deformation of any substances., e.g., in the deformation of amorphous or semi-crystalline polymers. Rubber elasticity is also essential to the functions of elastic proteins and muscles. Thus, the theory of rubber elasticity is centrally essential to much of polymer science. In this chapter, we have touched upon some of the basic concepts of thermodynamics of rubber elasticity and other factors affecting it.

Keywords: Thermodynamics, Elasticity, Phenomenological Treatment, rubber elasticity, rubber

1. Introduction

The capability of Rubber-like materials to extend to several-fold their original length is undoubtedly the most striking characteristic, which has been the subject of research interest for decades. Questions like what factor contributes to such substantial deformation and what effects temperature and pressure cast on the elasticity of rubber, all these have been pursued by several investigators. Having these questions in the back of our mind, we shall try to cement our concepts regarding how rubber elasticity is different from those of crystalline solid [1, 2] and glasses [3] which cannot normally be extended to more than a small fraction of their original length without undergoing failure, and ductile materials such as metals [4–6] which can undergo large deformation but cannot return to their original length upon removal of stress.

In this chapter, we will first get up to understanding the thermodynamics of rubber elasticity. It should be noted that the classical thermodynamic approach is only concerned with the macroscopic behavior of material under investigation and has very little thing to do with their molecular structure. The next section of this chapter presents the quantitative description of elasticity of the network of rubber chains based on the classical principles of statistical mechanics. Finally, we discuss various factors affecting the elasticity of rubber.

2. Effect of various factors on the elasticity of rubber

2.1 Effect of temperature on rubber elasticity

We need an experimental setup for investigating the effect of temperature on rubber elasticity. There are numerous ways to measure this experimentally.

One such experimental design that is widely used to probe the fundamentals of rubber elasticity is mentioned in **Table 1**. This experiment is primarily based on the automatic stress-relaxometry setup as shown in **Figure 1**, which was used by M.C. Shen and D.A. McQuarrie [7].

An outcome of such experimental exercise on natural rubber samples is shown in **Figure 2**, which exhibits the effect of temperature on the restoring force at various extension ratios. The testing sample was prepared from NBS pale creep rubber, which was cured by 1.5 phr dicumyl peroxide at 145°C temperature for 40 min duration. The restoring force is seemed linearly varying over a wide range of temperatures. The slop of the force-temperature curve however changes depending upon the extension ratio. The shift from negative slope at low degree of elongation to the positive slope at a higher extension ratio is called the *thermoelastic inversion*. The thermoelastic inversion value may lies around 10% for rubbery materials. It should be noted that this behavior is not confined to natural rubber, but it is general for all rubbery materials.

Now is the right time to get fully involved with the constitutive relationship between the restoring force (f) and various thermodynamic state variables such as temperature (T), pressure (P), etc. Most thermodynamic theories in general

Step 1. Extend the rubber sample to a desired fixed length	**Step 2.** Allow the sample to equilibrate, i.e., allowing the stress-relaxation to proceed until constant modulus is achieved.
Step 3. Under equilibrium conditions, measure the restoring force as a function of temperature at constant pressure.	**Step 4.** Repeat steps 1 to 3 on a fresh sample at different elongated lengths.

Table 1.
Procedure for measuring the effect of temperature on the elasticity of rubber.

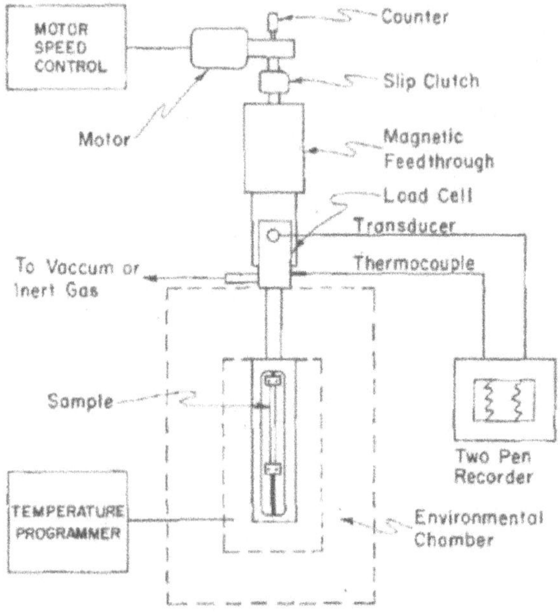

Figure 1.
The automatic stress relaxometry. By permission of the American Institute of Physics [7].

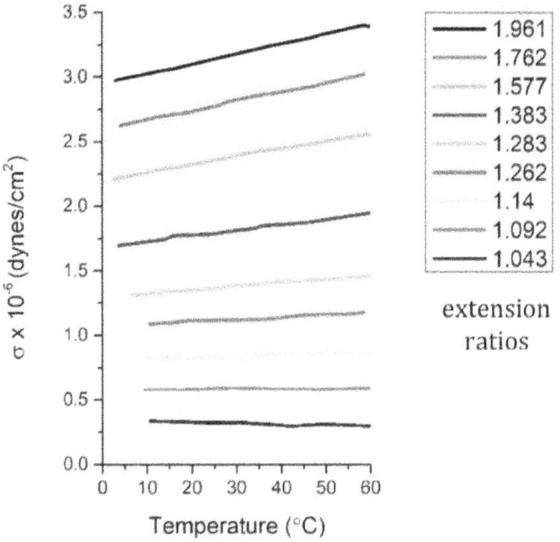

Figure 2.
Stress-temperature curves of natural rubber [7].

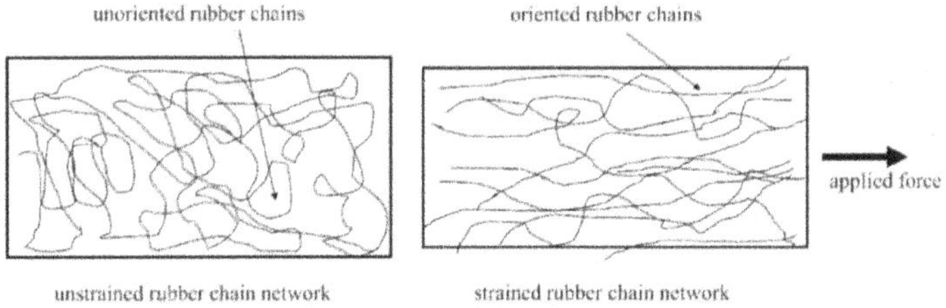

Figure 3.
Schematic representation of change in configuration of rubber chains under the applied stress.

textbooks are confined to the gaseous pressure-volume form of work ($dW = PdV$). However, in the case of the deformation of rubber, there is something more than just pressure-volume work. When a strip of rubber sample is elongated by a length dl, a restoring force is generated inside the rubber system (**Figure 3**). Therefore, the tiny amount of work involved in this process can be expressed as,

$$dW = PdV - fdl \tag{1}$$

In the expression above, the dV factor is the volume change that arises due to the elongation of the rubber sample. Usually, PdV is so small relative to the fdl that PdV can be dropped out from Eq. (1). However, we are going to have PdV for the sake of completeness. Now, for the reversible processes, we may combine the first and second law of thermodynamics and write it as,

$$dU = TdS - dW \tag{2}$$

From Eqs. (1) and (2)

$$dU = TdS - PdV + fdl \tag{3}$$

Since the experiment is usually conducted at a constant atmospheric pressure value (1 atm or 14.696 psi), the enthalpy transfer dH accompanying the volume change due to the elongation of the rubber may be written as,

$$dH = dU + PdV \tag{4}$$

By combining Eqs. (3) and (4), we arrive at the expression,

$$dH = TdS + fdl \tag{5}$$

By partially differentiating Eq. (5) and treating temperature and pressure as constant, we get,

$$f = \left(\frac{\partial H}{\partial l}\right)_{T,P} - T\left(\frac{\partial S}{\partial l}\right)_{T,P} \tag{6}$$

$$f = \left(\frac{\partial H}{\partial l}\right)_{T,P} + T\left(\frac{\partial f}{\partial T}\right)_{l,P} \tag{7}$$

Eqs. (6) and (7) hold an essential piece of information regarding the origin of elastic force. Let us take a moment to behold this relationship between restoring force and temperature and what role enthalpy and entropy play in this Eq. (6). According to Eq. (6), the restoring force depends on two factors: enthalpy and entropy change that occur in rubber due to elongation (or deformation in general terms). Generally, rubber molecules are so long that almost every chain participates in crosslinking and entanglement processes. During deformation or stretching, some rubber chains are forced to become linearly oriented, which causes a decrease in entropy of the rubber system. This decrease in entropy gives rise to the elastic force in the network chains, (**Figure 4**).

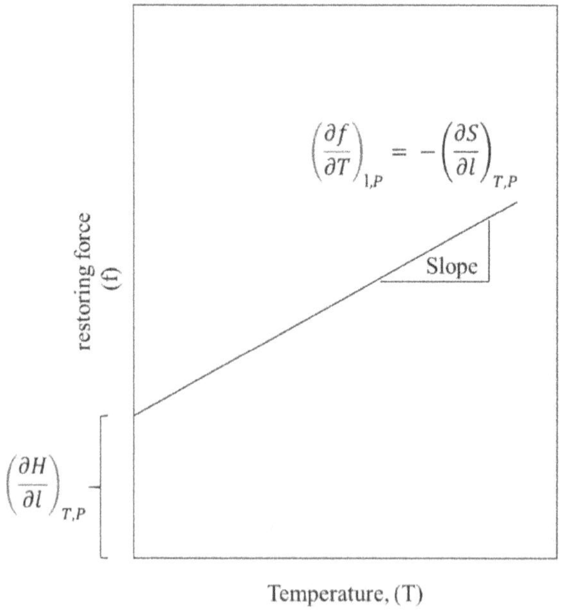

Figure 4.
Graphical representation of the relationship between restoring force and temperature.

It is important to understand the limitation of Eq. (6) that it does not fully comply with the behavior of rubber at extremely high elongation. The $\left(\frac{\partial H}{\partial l}\right)_{T,P}$ part has a finite value and cannot be ignored. At sufficiently high elongation, most rubber crystalizes and thus $\left(\frac{\partial H}{\partial l}\right)_{T,P}$ factor may overweight $-T\left(\frac{\partial S}{\partial l}\right)_{T,P}$. The coefficient $\left(\frac{\partial H}{\partial l}\right)_{T,P}$ can be experimentally obtained from the force-temperature curve as the intercept on restoring force axis at zero temperature value. The coefficient $\left(\frac{\partial H}{\partial l}\right)_{T,P}$ has a fundamental relationship with other thermodynamic quantities, which may be expressed as

$$\left(\frac{\partial H}{\partial l}\right)_{T,P} = \left(\frac{\partial U}{\partial l}\right)_{T,V} + T\frac{\alpha}{\beta}\left(\frac{\partial V}{\partial l}\right)_{T,P} \tag{8}$$

where, α is the cubical coefficient of thermal expansion,

$$\alpha = \frac{1}{V}\left(\frac{\partial V}{\partial T}\right)_{P,l} \tag{9}$$

and β is the coefficient of isothermal compressibility,

$$\beta = \frac{1}{V}\left(\frac{\partial V}{\partial P}\right)_{T,l} \tag{10}$$

The coefficient value α and β can be measured experimentally, however the coefficient $\left(\frac{\partial V}{\partial l}\right)_{T,P}$ is usually exceedingly small for most of the rubbers. Therefore, a great deal of experimental accuracy is required. Eqs. (7) and (8) can be combined to one the more refined and insightful expressions about the relative contributions of enthalpy and entropy towards the rubber elasticity. It should be noted that Eq. (11), in practice, does not comply well with the real behavior of rubber-like materials due to a lack of accurate $\left(\frac{\partial V}{\partial l}\right)_{T,P}$ data [8–10]. Therefore, various approximations have been suggested, which is beyond the scope of this chapter.

$$f = \underbrace{\left(\frac{\partial U}{\partial l}\right)_{T,V} + T\frac{\alpha}{\beta}\left(\frac{\partial V}{\partial l}\right)_{T,P}}_{\text{enthalpy contribution}} + \underbrace{T\left(\frac{\partial f}{\partial T}\right)_{l,P}}_{\text{entropy contribution}} \tag{11}$$

2.2 Effect of crosslinking on rubber elasticity

As discussed in the introduction section, rubber materials have the outstanding ability to return to their initial position almost instantaneously upon the removal of deforming load with virtually zero permanent deformation within the network chain structure. This snapping of rubber primarily happens due to the presence of crosslinks. We can think of crosslinks as the knot holding two or more threads together. Crosslink inhibits the long-range mobility of rubber chains. Therefore, based on our general understanding, we can confidently say that the crosslinked rubber would require a lot more force to stretch for the same amount of strain as the uncrosslinked rubber [11–13]. **Figure 5** schematically represents the crosslinking process of the linear polymer chains into an infinite network. In practice, crosslinking process is performed by incorporating the appropriate crosslinking agents like sulfur or peroxide into the rubber matrix and heating it at elevated temperature under some pressure. The crosslinking process enhances dimensional stability, abrasion resistance, and many other properties [14–17].

The extent to which a rubber chain network is crosslinked directly impacts the elasticity of rubber. According to the statistical mechanics, rubber elastic modulus under a uniaxial elongation is directly proportional to the number of crosslinks per unit volume (N_o or mol/cm^3) of the rubber chain network. If the rubber density is given as d (g/cm^3), and the average molecular weight of network chain is M_c (g/mol), then

$$N_o = \frac{d}{M_c} \tag{12}$$

For a small uniaxial strain value, the relationship between the elastic modulus E_o and crosslink concentration N_o can be mathematically written as

$$E_o = 3N_oRT\frac{\overline{\overline{r}}_o^2}{r_f^2} \tag{13}$$

or

$$E_o = \frac{3dRT}{M_c}\frac{\overline{\overline{r}}_o^2}{\overline{r}_f^2} \tag{14}$$

Here, $\overline{\overline{r}}_o^2$ represents the mean square end-to-end distance of the chains within the network, and $\overline{r}_f^2$ represents the mean end-to-end distance of the isolated chains [18, 19].

Since rubber is considered an incompressible material in relation to their shear deformation, that is poisson's ratio of rubber is close to 0.5, i.e., elastic modulus E_o is approximately equal to three times the shear modulus G_o. Therefore, Eq. (14) can be rewritten as,

$$G_o = \frac{dRT}{M_c}\frac{\overline{\overline{r}}_o^2}{r_f^2} \tag{15}$$

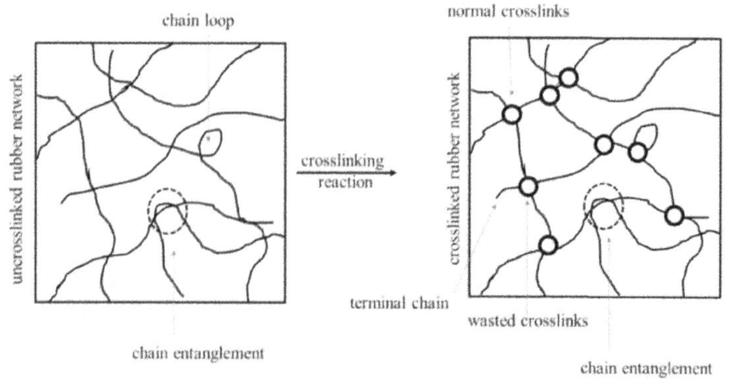

Figure 5.
Schematic representation of crosslinks in an ideal chain network structure.

Eq. (15), however is based on the idealized image of a perfect rubber chain network in which all network chains contribute to the elasticity of rubber. It is assumed that each crosslink combines four chains, and two crosslinks terminate each such chain. But in reality, there are several imperfections present in the rubber chain network. As illustrated in **Figure 5**, the non-effective crosslinks like wasted crosslinked, chain loops, and terminal ends significantly affect the elastic stress in the strained network of rubber chains. For example, linear polymer chains of average molecular weight M would have $2d/M$ number of terminals. Since these terminals would not take part in the elasticity, they should be excluded from the number of effective chains.

$$G_o = RT\frac{\bar{\bar{r}}_o^2}{r_f^2}\left(\frac{d}{M_c} - \frac{2d}{M}\right) \tag{16}$$

or,

$$G_o = \left(\frac{dRT}{M_c}\right)\frac{\bar{\bar{r}}_o^2}{r_f^2}\left(1 - \frac{2M_c}{M}\right) \tag{17}$$

Entanglements in rubber chain network also impose additional restriction, which leads to increment in the elastic stress. In a closed packed network of rubber chains, it is quite natural to expect several such entanglements between two consecutive crosslinks [20, 21]. Therefore, the contribution of such entanglements to elastic stress cannot be overlooked, especially chains that are long enough to permit multiple entanglements. The deviation from the "normal" crosslinked structure can be accounted for by adding the *entanglement factor* (a) in Eq. (18),

$$G_o = \left(\frac{dRT}{M_c} + a\right)\frac{\bar{\bar{r}}_o^2}{r_f^2}\left(1 - \frac{2M_c}{M}\right) \tag{18}$$

2.3 Effect of filler on rubber elasticity

Raw rubber is a mechanically weak material. Thus, rubber needs some extra compounding ingredients to enhance its physical properties in addition to crosslinking. Therefore, filler is an indispensable ingredient in the rubber industry. Carbon black, zinc oxide, silica, clay are some commonly used filler examples. Filler can be of two types, reinforcing and non-reinforcing. Reinforcing fillers increase the rubber's stiffness without impairing the strength and losing the rubbery characteristic.

The most common expression describing the effect of filler on rubber elasticity is popularly called as Guth-Smallwood equation

$$\frac{E_f}{E_o} = 1 + 2.5 * \varnothing_f + 14.1 * \varnothing_f^2 \tag{19}$$

where, $\varnothing_f$ is the volume fraction of filler and subscripts f and 0 refer to the filled and unfilled rubber respectively. Eq. (19) is inspired by the Einstein's equation that relate viscosity of fluid containing small solid suspension particles [22, 23]. The validation of Eq. (19) is to be found in its good agreement with the experimental data as shown in **Figure 6**.

Stress softening is another exciting behavior that is commonly seen in the filled rubber system. It was first observed by the Mullins, after whom it is named [24]. According to the stress-softening effect, the stress–strain curve depends on the maximum loading previously encountered. The term "Mullins effect" is also common to all rubbers, including non-filled rubber. **Figure** 7 illustrates the softening of stress with each step.

Bueche gave the first molecular interpretation of the Mullin's effect and the explanation for the **Figure** 7 [25]. As illustrated in **Figure** 8, two nearby filler particles in a reinforced rubber are connected via polymer chains. One of them is relaxed, but the others are relatively elongated. When the rubber sample is stretched, the "prestrained" chains first reach maximum extension, and then they either become detached or break. In the second cycle, the detached/broken chains no longer share the overall applied load, thus giving rise to the observed softening of stress behavior. The same thing happens when the sample is stretched a third time.

2.4 Effect of stress-induced crystallization

In unstrained conditions, the rubber sample is assumed to hold isotropic properties. However, when the rubber sample is stretched, anisotropic change occurs at

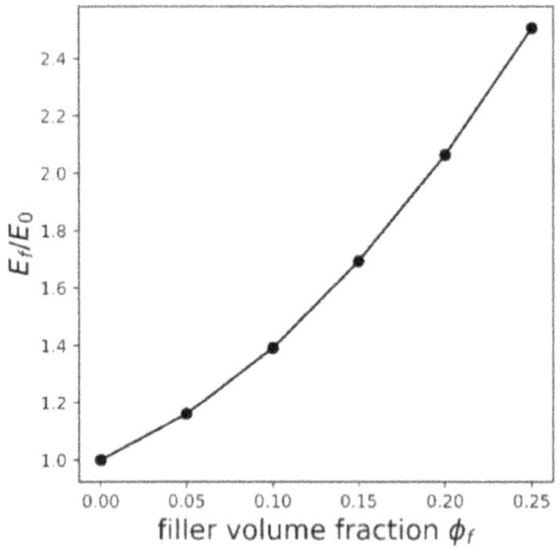

Figure 6.
Effect of filler volume fraction on filled to non-filled rubber modulus ratio.

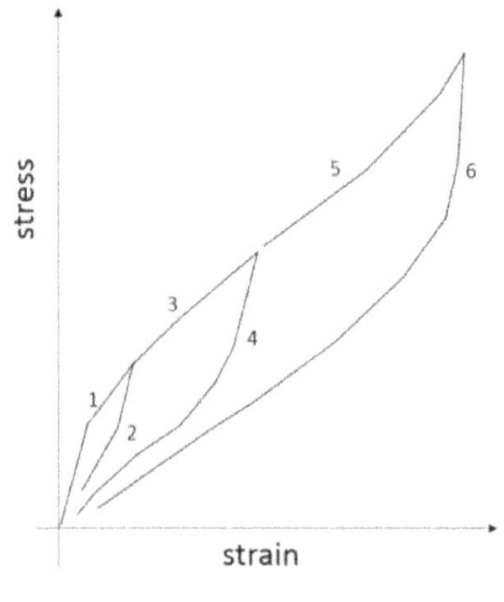

Figure 7.
Stress–strain curves for a filled rubber showing progressive cyclic softening, also known as the Mullins effect.

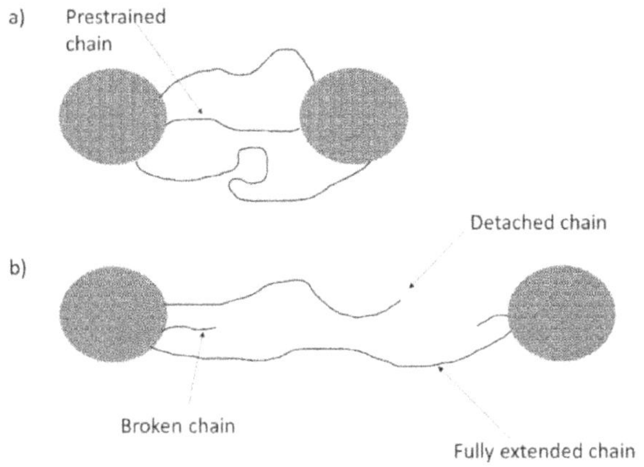

Figure 8.
Schematic representation of molecular mechanism for stress-softening effect. (a) Unstrained rubber chain-filler particle assembly, and (b) rubber chain-filler particle assembly after deformation.

the microscopic level. Polymer chains tend to orient more in the direction of stretch than in the lateral directions. Therefore, a greater number of ordered chins favor the formation of crystallites. These crystallites combine numbers of nearby network chains, which then act as crosslinks. As the sample is stretched further, more crystallite is formed and get transformed into physical crosslinks. These crosslinks then, in turn, cause a rise in elastic stress. It is known that such crystallites have very high elastic moduli ($\sim 10^{11}$ Pa), which is usually five orders of magnitude higher than the elastic moduli of rubbery materials ($\sim 10^{6}$ Pa). At higher elongation, such crystallites also play the role of reinforcing filler, which further increases the elastic stress of the rubber sample.

Figure 9 illustrates the stress–strain behavior of natural rubber carried at two temperatures, i.e., 30 and 60°C. The graph shows how stress steeply rises above a certain ratio, i.e., $\lambda = 0.3$. However, it should be noted that the stress-induced behavior also depends on the temperature at which experiment is conducted. At 60°C natural rubber sample exhibit slight upturn in the elastic stress towards the extension ratio, which is primarily caused by the finite extensibility of the polymer chain in the network [27].

2.5 Time-temperature superposition principle

So far, we have done a quantitative and qualitative discussion on the influence of various factors like temperature, extension ratio, crosslink density, fillers, and crystallinity affecting the elastic stress of modulus of the rubber sample. These considerations present a fair picture of physical behavior of a rubber material. However, the constitutive relations those we have seen till now are based on the static experimental data that is the effect of an immediate change to a system is calculated without regard to the longer-term response of the system to that change.

Rubber is regarded as a highly viscoelastic material that is rubber resembles characteristics of both elastic and viscous material. There are three main characteristics of viscoelastic materials: creep, stress relaxation, and hysteresis. These viscoelastic phenomena arise due to the long-term response of the material to a constant load or strain. Clearly, there is a time factor involved in these observations, which leads to the question that how one can relate viscoelastic responses with time scale mathematically. There are quite some constitutive models that quantitatively express viscoelastic response as a function of time. Maxwell, Kelvin–Voigt,

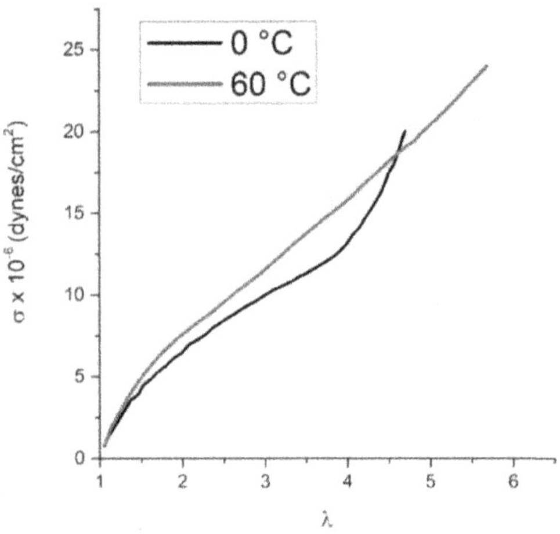

Figure 9.
Natural rubber stress–strain curve measured at two different temperatures [26].

Generalized Maxwell models are some well-known examples. Detailed discussion on these models is however beyond the scope of this chapter. Rather, we shall focus on understanding the time-dependent variation of elastic modulus by using time–temperature superposition principle.

To better understand time–temperature superposition principle let us us take an example of a stress relaxation experiment conducted at different temperature. At a given temperature T_1, a polymer sample of unit cross section area is subjected to an instantaneous strain that is maintained constant throughout the whole experiment. Then the stress as function of time is measured and stress relaxation modulus is obtained according to the Eq. (20). Where t can be experimentally accessible time period. Polymer sample is then set free from stress and allowed to undergo relaxation. Next, temperature is changes to T_2, and same procedure is repeated yielding $E(t)$tensile stress relaxation modulus at new T_2 temperature. Theis process is repeated at several different temperature and "*t second stress-relaxation modulus*" is obtained as a function of temperature.

$$E(t) = \frac{\sigma(t)}{\epsilon_0} \tag{20}$$

where, $\sigma(t)$ is stress as a function of time, and ϵ_0 is constant strain value.

Time–temperature superposition principle states that the change in temperature from T_1 to T_2 is equivalent to multiplying the time scale by a constant factor a_T that is only a function of the two temperatures T_1 to T_2 according to the Eq. (21).

$$E(t_1, T_1) = E\left({}^{t_1}\!/_{a_T}, T_2\right) \tag{21}$$

$$log\ a_T = log\ \frac{t_1}{t_2} \tag{22}$$

where t_2 is the time required to reach $E(t_2, T_2)$ stress relaxation modulus measured at T_2 temperature. **Figure 10(left)**. exhibits data obtained from one such experiment (i.e., stress relaxation) conducted on *bis*-phenol-A-polycarbonate (M_w = 40,000 g/mol), and **Figure 10(right)** represents the transformed modulus-time curve for a referenced temperature of 141°C [26].

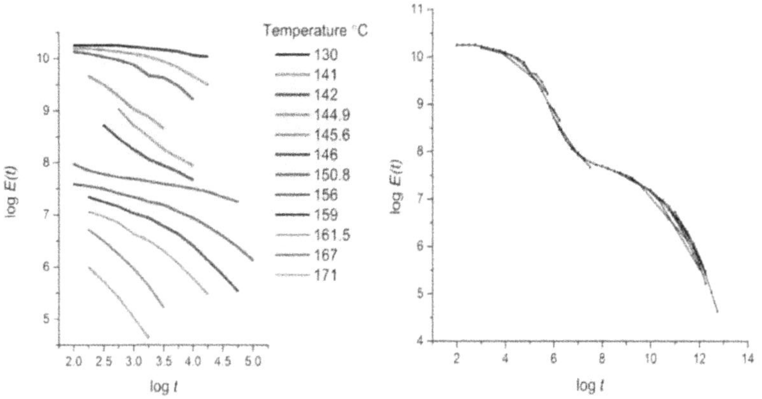

Figure 10.
Transformation of stress relaxation modulus-time obtained at different temperatures into single master curve for 140°C reference temperature.

3. Conclusions

Overall, we have discussed the basic concepts of thermodynamics of rubber elasticity. The quantitative effect of temperature on the elasticity of rubber has been scribbled down. Then the qualitative discussion on the role of crosslink density, filler concentration and strain induced crystallization on the elastic modulus of rubber-like materials provide supporting explanations backed by mechanism from statistical to molecular scale standpoint. Additionally, we discussed what role time scale plays in elastic modulus of general polymer at a given temperature using time–temperature superposition principle. The contents discussed in this chapter is simple and sufficient to develop an interest in readers' mind to take next step towards studying more complex rubbery materials.

Author details

Sanjay Pal*, Mithun Das and Kinsuk Naskar
Rubber Technology Centre, Indian Institute of Technology Kharagpur, India

*Address all correspondence to: sanjay.cusat.psrt@gmail.com

References

[1] G. Biroli and P. Urbani, "Breakdown of elasticity in amorphous solids," 2016, doi: 10.1038/NPHYS3845.

[2] "Are amorphous solids elastic or plastic? | EurekAlert!" https://www.eurekalert.org/news-releases/498690 (accessed Aug. 23, 2021).

[3] D. R. (Donald R. Uhlmann and N. J. Kreidl, "Elasticity and strength in glasses," p. 282, 1980.

[4] D. François, A. (André) Pineau, and A. Zaoui, "Mechanical Behaviour of Materials : Volume I: Elasticity and Plasticity," p. 440, 1998.

[5] W.-F. Chen and A. F. (Atef F. Saleeb, "Constitutive equations for engineering materials," p. 1129, 1994.

[6] A. C. Ugural and S. K. Fenster, "Advanced mechanics of materials and applied elasticity," p. 680, 2012.

[7] M. C. Shen, D. A. McQuarrie, and J. L. Jackson, "Thermoelastic Behavior of Natural Rubber," *J. Appl. Phys.*, vol. 38, no. 2, p. 791, Dec. 2008, doi: 10.1063/1.1709414.

[8] R. W. Ogden, "Volume changes associated with the deformation of rubber-like solids," J. Mech. Phys. Solids, vol. 24, no. 6, pp. 323–338, Dec. 1976, doi: 10.1016/0022-5096(76)90007-7.

[9] G. Gee, J. Stern, and L. R. G. Treloar, "Volume changes in the stretching of vulcanized natural rubber," Trans. Faraday Soc., vol. 46, no. 0, pp. 1101–1106, Jan. 1950, doi: 10.1039/TF9504601101.

[10] F. G. Hewitt and R. L. Anthony, "Measurement of the Isothermal Volume Dilation Accompanying the Unilateral Extension of Rubber," *J. Appl. Phys.*, vol. 29, no. 10, p. 1411, Jun. 2004, doi: 10.1063/1.1722959.

[11] L. H. Sperling, "INTRODUCTION TO PHYSICAL POLYMER SCIENCE FOURTH EDITION," 2006, Accessed: Aug. 12, 2021. [Online]. Available: www.wiley.com.

[12] H. Mark, "Some Applications of the Kinetic Theory to the Behavior of Long Chain Compounds," *J. Appl. Phys.*, vol. 12, no. 1, p. 41, Apr. 2004, doi: 10.1063/1.1712850.

[13] P. J. Flory and J. R. Jr., "Statistical Mechanics of Cross-Linked Polymer Networks II. Swelling," *J. Chem. Phys.*, vol. 11, no. 11, p. 521, Dec. 2004, doi: 10.1063/1.1723792.

[14] J. D. Ferry, "Viscoelastic properties of polymers, 3rd edition," *Wiley, New York*, 1980.

[15] M. L. Huggins, "Properties and structure of polymers, A. T. TOBOLSKY. Wiley, New York, 1960, IX + 331 pp. $14.50," J. Polym. Sci., vol. 47, no. 149, pp. 537–537, Nov. 1960, doi: 10.1002/POL.1960.1204714974.

[16] J. P. Mercier, J. J. Aklonis, M. Litt, and A. V. Tobolsky, "Viscoelastic behavior of the polycarbonate of bisphenol A," J. Appl. Polym. Sci., vol. 9, no. 2, pp. 447–459, Feb. 1965, doi: 10.1002/APP.1965.070090206.

[17] H. M. James and E. Guth, "Theory of the Elastic Properties of Rubber," *J. Chem. Phys.*, vol. 11, no. 10, p. 455, Dec. 2004, doi: 10.1063/1.1723785.

[18] L. R. G. Treloar, "Stress-strain data for vulcanised rubber under various types of deformation," Trans. Faraday Soc., vol. 40, no. 0, pp. 59–70, Jan. 1944, doi: 10.1039/TF9444000059.

[19] U. W. Suter and P. J. Flory, "Conformational Energy and Configurational Statistics of Polypropylene," Macromolecules,

vol. 8, no. 6, pp. 765–776, Nov. 2002, doi: 10.1021/MA60048A018.

[20] P. J. Flory, "Theoretical predictions on the configurations of polymer chains in the amorphous state," http://dx.doi.org/10.1080/00222347608215169, vol. 12, no. 1, pp. 1–11, Jan. 2008, doi: 10.1080/00222347608215169.

[21] P. J. Flory, "Statistical thermodynamics of random networks," Proc. R. Soc. London. A. Math. Phys. Sci., vol. 351, no. 1666, pp. 351–380, Nov. 1976, doi: 10.1098/RSPA.1976.0146.

[22] H. M. Smallwood, "Limiting Law of the Reinforcement of Rubber," *J. Appl. Phys.*, vol. 15, no. 11, p. 758, Apr. 2004, doi: 10.1063/1.1707385.

[23] A. M. Bueche, "A Physical Theory of Rubber Reinforcement," *J. Appl. Phys.*, vol. 23, no. 1, p. 154, Jun. 2004, doi: 10.1063/1.1701968.

[24] L. Mullins, "Determination of degree of crosslinking in natural rubber vulcanizates. Part IV. Stress-strain behavior at large extensions," J. Appl. Polym. Sci., vol. 2, no. 6, pp. 257–263, Nov. 1959, doi: 10.1002/APP.1959.070020601.

[25] F. Bueche, "Molecular basis for the mullins effect," J. Appl. Polym. Sci., vol. 4, no. 10, pp. 107–114, Jul. 1960, doi: 10.1002/APP.1960.070041017.

[26] J. P. Mercier, J. J. Aklonis, M. Litt, and A. V. Tobolsky, "Viscoelastic behavior of the polycarbonate of bisphenol A," J. Appl. Polym. Sci., vol. 9, no. 2, pp. 447–459, Feb. 1965, doi: 10.1002/APP.1965.070090206.

[27] J. J. Aklonis and W. J. MacKnight, "Introduction to polymer viscoelasticity," p. 295, 1983.

Chapter 2

Nanostructures Failures and Fully Atomistic Molecular Dynamics Simulations

José Moreira de Sousa

Abstract

Nowadays, the concern about the limitations of space and natural resources has driven the motivation for the development of increasingly smaller, more efficient, and energy-saving electromechanical devices. Since the revolution of "microchips", during the second half of the twentieth century, besides the production of micro-computers, it has been possible to develop new technologies in the areas of mechanization, transportation, telecommunications, among others. However, much room for significant improvements in factors as shorter computational processing time, lower energy consumption in the same kind of work, more efficiency in energy storage, more reliable sensors, and better miniaturization of electronic devices. In particular, nanotechnology based on carbon has received continuous attention in the world's scientific scenario. The riches found in different physical properties of the nanostructures as, carbon nanotubes (CNTs), graphene, and other exotic allotropic forms deriving from carbon. Thus, through classical molecular dynamics (CMD) methods with the use of reactive interatomic potentials reactive force field (ReaxFF), the scientific research conducted through this chapter aims to study the nanostructural, dynamic and elastic properties of nanostructured systems such as graphene single layer and conventional carbon nanotube (CNTs).

Keywords: molecular dynamics method, interatomic reactive force field—ReaxFF, graphene monolayer, convetional carbon nanotubes—(CNTs), elastic properties

1. Introduction

Failures in condensed materials can be observed from the naked eye, Earth's crust in earthquakes, to the interatomic interaction of atoms and molecules at the nanoscale (nanoscience), not visible in experimental procedures that require explanations of certain physical phenomena on the nanometric scale [1].

Thus, understanding the condensed matter under mechanical load is of fundamental importance in the sustained development of new materials with superior qualities to the existing ones. Understanding the structural failure (even at the nanometric scale), the mechanical limits of new materials allow us to establish nanostructure applications and new materials for certain purposes the applicability in the development of new electromechanical devices to specific applicability's in advance nanotechnologies. Over several thousands of years, knowing the behavior of condensed matter under extreme conditions of mechanical stress has provided the way for a new era in the science of materials and its modern technologies for improving the quality of life for humans and planet Earth. Due to nanotechnological advances, the human quality of life has important improvements in material science and its technologies. So, the basis of this advance is in the study of the physical properties of materials and nanostructures (theoretically as well as experimentally) and their various length and time scales for theoretical study for physicochemical predictions of nanostructures and materials. Fully atomistic molecular dynamics, like computational modeling, is becoming increasingly important and indispensable in theoretical description and predictions not understandable by experimental scientists in the development of new technologies [2, 3]. Nevertheless, starting to create nanostructures at the scale of atoms and molecules, atomistic models are described in terms of length-scale computational cost to obtain theoretical results of the physical properties of nanostructures and new materials. The following is an illustration of the computational cost of fully atomistic modeling at the scale of atoms and molecules [4], together with its computational methods widely used in computational modeling in materials science [5] (see **Figure 1**).

A fundamental and very important concept in the study and analysis of mechanical failure in nanostructures and new materials is to establish valid methods obtained from experimental averages. Thus, it is possible to establish a fully atomistic computational modeling to model the physical properties of nanostructures and new materials, where the set of parameters described in the reactive and non-reactive force fields are obtained directly from the results provided by the experiments. Currently, the combination of experimental tests with computational modeling concepts has shown promising and efficient results in the study of the physical properties of nanostructures and new materials at accessible computational cost scales with the dimensions of the nanostructure (number of atoms and molecules) [6, 7]. This strategy in nanotechnology has contributed to important results in simulations of atoms and molecules in scientific and technological innovation and applications [4, 8]. Thus, simulations with carbon nanostructures have received particular attention after the synthesis of graphene, which won a Nobel Prize for K.S. Novoselov, A.K. Geim in 2004 [9] (by mechanical exfoliation method of graphite), thus suggesting a new era in materials science and fully atomistic

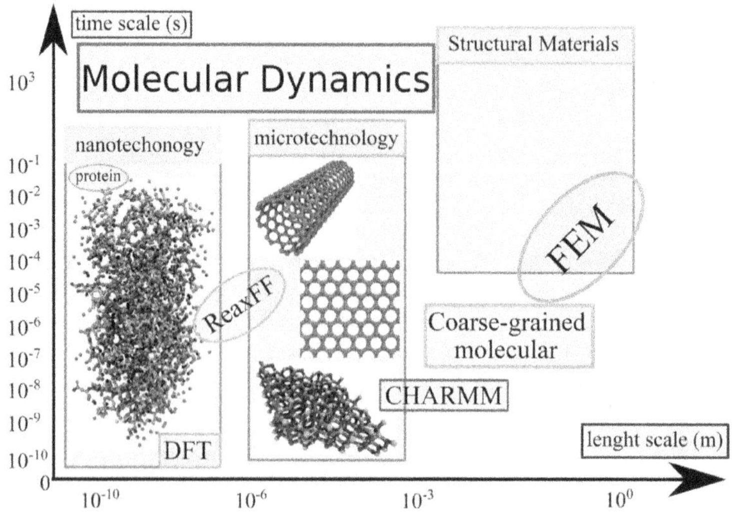

Figure 1.
Overview of a diagrammatic representation of timescales and lengthscale associated with computational methods used for computational simulation in the development of new materials in time and length scales [4].

computational simulations of systems formed by carbon atoms, particularly such as "graphene", a single layer of graphite and carbon nanotubes (CNTs), the cylindrical shape of roll-up one-dimensional graphene membrane (CNTs) [10].

In this chapter, we present the mechanical properties of graphene and CNT. We seek to show the efficiency of computational methods of reactive molecular dynamics with interatomic potentials parameterized by experimental results. We show theoretical results of mechanical failures in graphene monolayers and CNTs at the nanometer scale. Through computer simulations via classical molecular dynamics (CMD) method using the reactive force field (ReaxFF) reactive interatomic potential, we show mechanical failures (fracture pattern), Young's modulus, ultimate tensile strength (UTS), and critical strain for graphene and CNT and thus compare with the experimental results obtained in the literature. We hope that this chapter will add to future scientists who seek to start their academic activities using the molecular dynamics method with reactive potentials for studies not only of mechanical failures in nanostructures, but also more complete and detailed studies of the physical properties of nanostructured systems in nanometric scale.

2. Classical molecular dynamics simulation method

Classical molecular dynamics (CMD) is a technique that studies the behavior of a system of particles (atoms and molecules) as a function of time. The temporal evolution of set of these particles, in certain interacting systems, are obtained by the integration of equations of motion. Based on this, terms like "modeling" and "simulation" are widely used in conjunction with the numerical solution of physical problems involving interacting particles. However, it is important to note that these words have different meanings. The term "Modeling" refers to the development of the mathematical model of a physical situation, while "Simulation" refers to the procedure for solving equations, resulting in the developed model. This makes MD a widely used tool for studying material properties as an intersection of various scientific disciplines as shown in **Figure 2** [3].

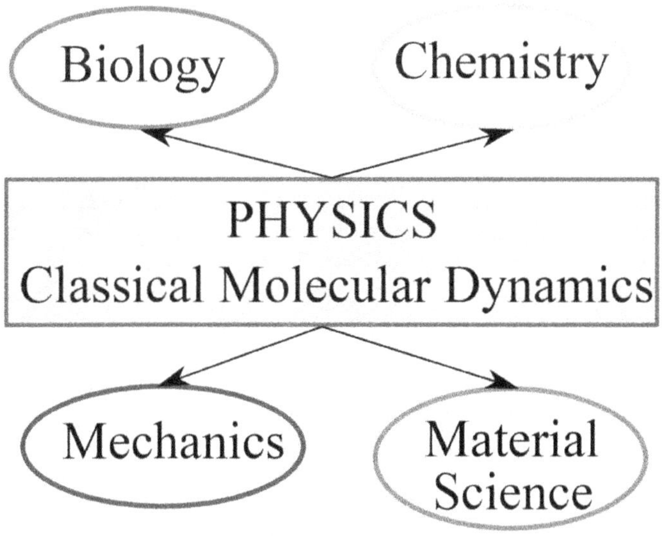

Figure 2.
Classical molecular dynamics method (CMD) as an intersection of several disciplines.

CMD simulations is a method that calculates the equilibrium and thermal transport properties in classical systems involving many bodies (in this case atoms as classical particles). In this context, the word "classic" means that the movement of these particles obeys the laws of classical mechanics (Newton's laws), being an excellent approximation for the study of the physical properties of a large number of nanostructures and new materials, especially graphene and CNT (in a study in this chapter). This method consists of solving Newton's equations for a set of atoms and molecules, thus obtaining the speed and position of each particle that makes up the physical system at each instant of the simulation. The theoretical basis of CMD embodies many of the results produced by great names in analytic mechanics such as: Euler, Hamilton, Lagrange, and Newton. Your contributions can be found in mechanics textbooks [11–13]. Some of these results contain fundamental observations of nature, while the others are elegant reformulations in the theoretical development of a classical mechanics set of linked computers that work together (computer cluster), perform computational calculations as a single system. The shared memory is performed by multi-threading parallelism (OpenMP) for computational clusters. Thus, after having the code installed and depending on what you want to simulate, it is necessary to build a computational code in C++ language, where we establish the physical properties of the problems to be studied by performing the computational modeling. For example, in this review chapter, we will simulate the mechanical failures of graphene and CNTs by the CMD-ReaxFF method. The code is written in computational language C++, where the thermodynamic quantities output via the "thermo-style" command is important to normalize all physics quantities by the number of atoms. This behavior can be changed via the thermo-modify (in real units) norm command. After the initial definitions of the code, we establish the statistical set that will describe the physical properties of the computational sample that we intend to computationally simulate, such as the NVT statistical canonical ensemble. In many cases, because the system has a very large number of particles, it is impossible to find the properties of such complex systems analytically. The trajectories of atoms and molecules are determined through from the numerical solution of Newton's equations of motion, to a system with interacting particles, where the force between the particles and the potential energy are defined by mechanics force field molecular (here reactive force field—ReaxFF discussed in following section).

Therefore, the objective in an atomistic simulation is to predict the movement of each atom in a material, characterized by a set of linked computers that work together (computer cluster), perform computational calculations as a single system. The shared memory is performed by multi-threading parallelism (OpenMP) for computational clusters. Thus, after having the code installed and depending on what you want to simulate, it is necessary to build a computational code in C++ language, where we establish the physical properties of the problems to be studied by performing the computational modeling. For example, in this review chapter, we will simulate the mechanical failures of graphene and CNTs by the CMD-ReaxFF method. The code is written in computational language C++, where the thermodynamic quantities output via the "thermo-style" command is important to normalize all physics quantities by the number of atoms. This behavior can be changed via the thermo-modify (in real units) norm command. After the initial definitions of the code, we establish the statistical set that will describe the physical properties of the computational sample that we intend to computationally simulate, such as the NVT statistical canonical ensemble: the atomic position $r_i(t)$, the velocities $v_i(t)$, and accelerations $a_i(t)$, as shown in **Figure 3.** The general idea of running a molecular dynamics simulation is presented by two factors as followed:

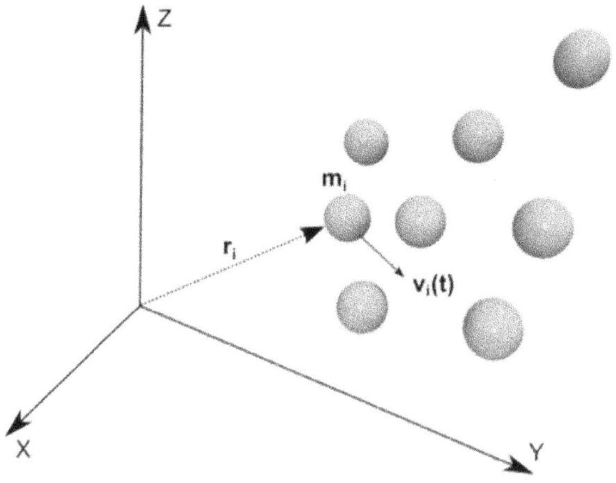

Figure 3.
Illustration of a system with N carbon atoms interacting with each other by interatomic reactive potentials, here in this chapter the ReaxFF reactive force field will be presented (see [4]).

1. Particle under the effect of potential energy, of where the forces governing the system can be calculated.

2. Equations of motion that determine the dynamics of particles, in which case Newton's laws are applied. Molecular dynamics uses Hamilton's classical equations of motion.

The classical Hamiltonian is defined as the sum of kinetic energy and energy potential:

$$p = -\frac{\partial \mathrm{H}}{\partial R_i} \tag{1}$$

and

$$\dot{R} = -\frac{\partial \mathrm{H}}{\partial p_i} \tag{2}$$

that lead to Newton's equations of motion. The classical Hamiltonian is defined as:

$$H(p_i, R_i) = \sum_{i=1}^{N} \frac{p_i^2}{2M_i} + V(R_i) \tag{3}$$

The force on an atom can be calculated by Newton's law as the derivative of energy in relation to the change in the position of the atom:

$$F_i = m_i \frac{d^2 Ri}{dt^2} = -\nabla_i V(R_i) = -\frac{dV}{dR_i} \tag{4}$$

leading to the set of Newtonian equations of motion for each particle i with mass m_i and Cartesian coordinate R_i. Therefore, for a closed system composed of N carbon atoms that interact through a potential energy function (here the interatomic reactive force field—ReaxFF), the CMD consists of solving the coupled N Newton equations. Therefore, in a computer simulation, we use a numerical integration algorithm to solve N differential equations [14].

In the classical formalism of CMD, the carbon atoms are treated as a collection of classical particles that can be described by Newtonian forces, where they are treated by harmonic or elastic forces. A complete set of interaction potentials between particles is known as the force field [15]. Parameters associated with force fields can be determined via first-principle calculations (Density Functional Theory—DFT) or via experimental results. Currently, there are numerous cam types of forces that are widely used in the study of the physical and chemical properties of nanostructured systems. We present in this chapter the interatomic force field—ReaxFF in the study of mechanical failure (mechanical properties) of a single layer of graphene and CNTs (armchair and zig-zag). In followed section, we show a brief description of the force field used in the simulations presented here in this chapter about the mechanical properties of graphene and CNT.

3. Interatomic reactive force field: ReaxFF

The reactive force field (ReaxFF) was developed to be a bridge between the chemical-quantum (QC) and the empirical (EFF) force fields [16, 17]. The EFF methods [18] describe the relationship between energy and geometry using a relatively simple set of functions. In the simplest form, EFF methods treat CMD systems or condensed matter systems by simple harmonic equations that describe the stretching and compression of bonds and the bending of bond angles. Unbound interactions are described by van der Waals potentials and Coulomb interactions (Lennard-Jones potential):

$$V_{LJ} = 4\in\left[\left(\frac{x}{\sigma}\right)^{-12} - \left(\frac{x}{\sigma}\right)^{-6}\right] \quad (5)$$

The Lennard-Jones potential consists of a two-body interaction function composed of the sum of two terms, an attractive interaction of the van der Waals type $\approx 10^{-6}$ and a short-range repulsive interaction $\approx 10^{-12}$ associated with the repulsion between orbitals atomic due to the Pauli exclusion principle. The terms # is a measure of the depth of the potential well and the term σ is the coefficient of the expression of the equilibrium distance of the pair of atoms. Classic models are not the only possible way to develop the potentials of many bodies. Developments based on first principles can lead to more accurate potentials for describing cases of interest. In this class of more modern potentials are included the so-called reactive potentials, developed specifically for a description of the dynamics of formation and breaking of bonds in materials. As reactive potential, we have the ReaxFF [16, 17], potential using in the chapter in the description of the physical properties of failure mechanics of graphene single layer and CNT. The CMD-ReaxFF is performed in all calculations in this chapter in the review study of mechanical properties of graphene single layer and CNTs. We show the interesting and important method, because through the theoretical results the values obtained in all simulations are in good agreement with experimental results and with results based on quantum methods (ab initio and DFT).

The modern reactive force field (ReaxFF) is parametrized whit first-principles calculations and compared with experimental results. The heath formations of the carbonous nanostructures values change between 2.8 and 2.9 kcal/mol when compare reactive molecular dynamics simulations and data experimental [16, 17]. The set parameter validity used for performed reactivity between carbon atoms ligands in this reviewer is divided by partial energy contributions [16, 17]:

$$E_{system} = E_{bond} + E_{over} + E_{under} + E_{val} + E_{pen} + E_{tor} + E_{conj} + EvdW + E_{\infty}, \quad (6)$$

where, here the terms of Eq. (6), respectively, represents the energies corresponding to the bond distance (E_{bond}), the over-coordination (E_{over}), the undercoordination (E_{under}), the valence (E_{val}), the penalty for handling atoms with two double bonds (E_{pen}), the torsion (E_{tor}), the conjugated bond energies (E_{conj}), the van der Waals ($EvdW$), and coulomb interactions (E_{co}). The fundametation of ReaxFF is bond order BO'_{ij} between a pair of atoms as [16, 17]:

$$BO'ij = \exp\left[pbo,1.\left(\frac{r_{ij}}{r_o}\right)^{pbo,2}\right] + \exp\left[pbo,3.\left(\frac{r_{ij}^{\pi}}{r_o}\right)^{pbo,4}\right] + \exp\left[pbo,5.\left(\frac{r_{ij}^{\pi}}{r_o}\right)^{pbo,6}\right] \quad (7)$$

where the atomic configurations is obtained from interatomic distance r_{ij} of three exponential terms, such as, the σ bond $(pbo, 1)$ and $(pbo, 2)$, first π bond $(pbo, 3)$ and $(pbo, 4)$ and $\pi\pi$ bond $(pbo, 5)$ and $(pbo, 6)$, with their respective dependencies in interatomic distances C—C bond $(\sigma \sim 1.5A)$, $(\pi \sim 1.2A)$ and $(\pi\pi \sim 1.0A)$.

The bond order is corrected for cases where there is over-coordination (more bonds than allowed), through f_1 and residual link order BO' for valence angles, through f_4 and f_5. The correction due to over-coordination occurs only for bonds between two carbon atoms, while the correction for the residual bond order BO' for valence angles occurs for all connections. The corrections f_i, $(f_i = 1...5)$ are presented in Eqs. (11)–(15). Bond order BO' for valence angle refers to the bond order existing between two atoms, not directly connected, where both are connected to a third, forming a valence angle [16, 17]:

$$\begin{aligned} BO_{ij} &= BO'_{ij}.f_1\left(\Delta'_i, \Delta'_j\right).f_4\left(\Delta'_i, BO'_{ij}\right).f_5\left(\Delta'_j, BO'_{ij}\right) \\ BO^{\sigma}_{ij} &= BO'^{\sigma}_{ij}.f_1\left(\Delta'_i, \Delta'_j\right).f_4\left(\Delta'_i, BO'_{ij}\right).f_5\left(\Delta'_j, BO'_{ij}\right) \\ BO^{\pi}_{ij} &= BO'^{\pi}_{ij}.f_1\left(\Delta'_i, \Delta'_j\right).f_1\left(\Delta'_i, \Delta'_j\right).f_4\left(\Delta'_i, BO'_{ij}\right).f_5\left(\Delta'_j, BO'_{ij}\right) \end{aligned} \quad (8)$$

$$BO^{\pi\pi}_{ij} = BO'^{\pi\pi}_{ij}.f_1\left(\Delta'_i, \Delta'_j\right).f_1\left(\Delta'_i, \Delta'_j\right).f_4\left(\Delta'_j, BO'_{ij}\right).f_5\left(\Delta'_j, BO'_{ij}\right) \quad (9)$$

$$BO_{ij} = BO^{\sigma}_{ij} + BO^{\pi}_{ij} + BO^{\pi\pi}_{ij} \quad (10)$$

$$f_1\left(\Delta'_i\Delta'_j\right) = \frac{1}{2}\left(\frac{Val_i + f_2\left(\Delta'_i\Delta'_j\right)}{Val_i + f_2\left(\Delta'_i\Delta'_j\right) + f_3\left(\Delta'_i\Delta'_j\right)} + \frac{Val_i + f_2\left(\Delta'_i\Delta'_j\right)}{Val_i + f_2\left(\Delta'_i\Delta'_j\right) + f_3\left(\Delta'_i\Delta'_j\right)}\right) \quad (11)$$

$$f_2\left(\Delta'_i\Delta'_j\right) = \exp\left(\lambda_1\Delta'_j\right) + \exp\left(-\lambda_1\Delta'_j\right) \quad (12)$$

$$f_3\left(\Delta'_i\Delta'_j\right) = \frac{1'}{\lambda_2}\ln\left\{\frac{1}{2}\left[\exp\left(-\lambda_2\Delta'_i\right) + \exp\left(-\lambda_2\Delta'_j\right)\right]\right\} \quad (13)$$

$$f_4\left(\Delta'_iBO'_{ij}\right) = \frac{1}{1 + \exp\left[-\lambda_3.\left(\lambda_4 BO'_{ij}BO'_{ij} - \Delta'_i\right) + \lambda_5\right]} \quad (14)$$

$$f_5\left(\Delta_i' BO_{ij}'\right) = \frac{1}{1 + \exp\left[-\lambda_3.\left(\lambda_4 BO_{ij}' BO_{ij}' - \Delta_i'\right) + \lambda_5\right]} \tag{15}$$

There are ReaxFF implementations, developed by individual research-based in [16, 17] formalism. Nowadays, the current ReaxFF parameter set developed by CMD-ReaxFF based on the periodic table of elements found on our planet Earth are [18, 19]:

- Group 1: H (non-metal), Li, Na, K, Rb, Cs (metals).
- Group 2: alkaline Earth metal: Mg, Ca, Sr and Ba.
- Group 3: lanthanide: Y.
- Group 4: lanthanide: Ti, Zr and Hf.
- Group 5: lanthanide: V and Nb.
- Group 6: lanthanide: Cr, Mo and W.
- Group 7: lanthanide: Mn.
- Group 8: lanthanide: Fe and Ru.
- Group 9: lanthanide: Co.
- Group 10: lanthanide: Ni, Pd and Pt.
- Group 11: lanthanide: Cu, Ag and Au.
- Group 12: lanthanide: Zn.
- Group 13: B, Al and Ga.
- Group 14: C and Si.
- Group 15: N and P.
- Group 16: O, S, Se and Te.
- Group 17: F, Cl and I.
- Group 18: He, Ne, Ar, Kr and Xe.

So, in this chapter, we performed reactive molecular dynamics simulations with ReaxFF to obtain the failure mechanics of a single layer of graphene and CNT. The results of reactive molecular dynamics simulations performed in this chapter (CMD-ReaxFF) of review are discussed in the next section.

4. Large-scale atomic/molecular massively parallel simulator code: LAMMPS

All simulations developed in this thesis were performed using the large-scale atomic/molecular massively parallel simulator (LAMMPS) code [20]. LAMMPS is a code that simulates a set of particles (solid, liquid or gas) using the classical molecular dynamics method. It is a code designed to obtain efficiency in the simulation when it is performed on parallel processors for systems whose particles are in a 3D rectangular box with density approximately uniform. It is an open source program maintained and distributed by researchers at Sandia National Laboratories [20], written in C++. It's a stable program that has the ability to simulate from a few particles to billions of them. In the following chapters we present the results that were obtained with the use with LAMMPS code.

4.1 Computational modeling of mechanical failure of nanostructures

LAMMPS code is a classical molecular dynamics simulation in language C++ designed to run efficiently on parallel computers. Is an open-source code, distributed freely under the terms of the GNU Public License (GPL) developed at Sandia National Laboratories [20]. The LAMMPS runs on a single processor or in parallel, or in single laptops or advanced computational clusters parallel using memory message-passing parallelism (MPI) [21].

A set of linked computers that work together (computer cluster), perform computational calculations as a single system. The shared memory is performed by multi-threading parallelism (OpenMP) for computational clusters. Thus, after having the code installed and depending on what you want to simulate, it is necessary to build a computational code in C++ language, where we establish the physical properties of the problems to be studied by performing the computational modeling. For example, in this review chapter, we will simulate the mechanical failures of graphene and CNTs by the CMD-ReaxFF method. The code is written in computational language C++, where the thermodynamic quantities output via the "*thermo-style*" command is important to normalize all physics quantities by the number of atoms. This behavior can be changed via the thermo-modify (in real units) norm command. After the initial definitions of the code, we establish the statistical set that will describe the physical properties of the computational sample that we intend to computationally simulate, such as the NVT statistical canonical ensemble:

$$\text{fix ID group} - \text{ID nvt temp } T_{\text{initial}}\ T_{\text{final}}\ T_{\text{damp}} \tag{16}$$

After the initial definitions of the code, we establish the statistical set that will describe the physical properties of the computational sample that we intend to computationally simulate, such as the NVT statistical ensemble. The NVT commands perform time integration on Nose-Hoover thermostat [22] style designed to generate positions and velocities of computational sampled under computational modeling by CMD-ReaxFF.

5. Canonical ensemble in statistical mechanics and thermodynamics quantities

Characterized by a set of macroscopic parameters, graphene, and CNT are in contact with the thermal reservoir (see **Figure 4**). Considering the very large reservoir compared to the computational sample that we intend to study, the total energy of the E_0 system, we have the validity of the thermodynamic postulates and statistical mechanics. Thus, the probability Pi of obtaining physical quantities of interest, in these cases, in the study of the mechanical failure of graphene and CNTs is given by [23]:

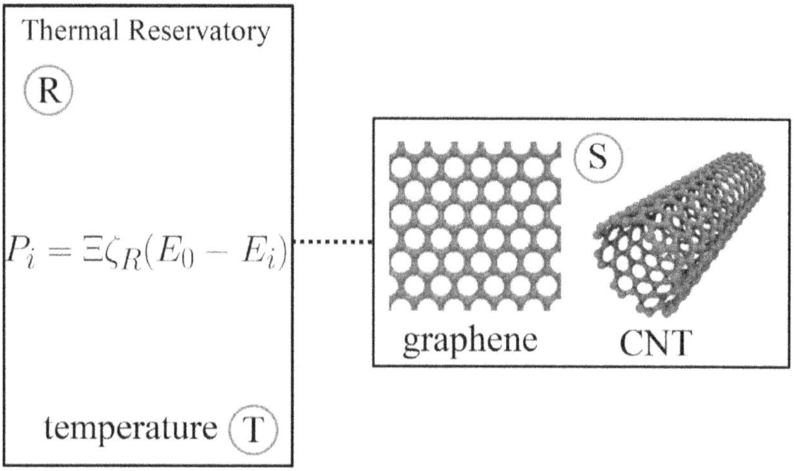

Figure 4.
Computational sample (graphene and CNT) in contact with a thermal reservoir R *at a specific temperature different at 0.*

$$P_i = \Xi\zeta_R(E_0 - E_i) \tag{17}$$

where Ξ is a normalization constant, E_i is the energy of the system in the particular thermodynamics state i, and ζ_R is a microscopic state accessible to the thermal reservoir with energy E.

After coupled canonical thermostat NVT in the graphene and CNT, we applied a constant engineering strain rate of $\delta = 10^{-6}\ \text{fs}^{-1}$, see de command line in LAMMPs code:

$$\text{fix ID group} - \text{ID deform N parameter args} \ldots \text{keyword value} \ldots \tag{18}$$

and so, adapt the code to the mechanical problem that seeks to study its elastic physical properties.

6. Fully atomistic computational simulation: elastic properties of graphene and CNT

After all the technical and physical properties in the study of mechanical failure in nanostructured systems, we present below the elastic properties of graphene single layer and CNTs (see **Figure 5**). In the **Figure 5**, we showed que atomistic configurations of graphene single layer 90 × 90°A whith 3256 carbon atoms, conventional carbon nanotubes whthi chirality armchair (11, 11) whith 616 carbon atoms and zig-zag (11, 0) whith 352 carbon atoms.

Thus, in **Figure 6**, we can see the graphical representations of the stress/strain curve for graphene monolayer at 300, 600, and 1000 K temperatures. For graphene monolayer at 300, 600 and, 1000 K (black, red, and blue curves), we clearly note two regimes: first, a linear regime followed by a plastic regime up to the complete fracture (see **Figure 6**). At 300 K, our results obtained by fully-atomistic reactive molecular dynamics simulations performed with interatomic force field ReaxFF we can see at room temperature a linear regime, we do not see permanent deformations which are different from what occurs for the plastic regime, where the graphene monolayer present breaking bonds (aligned in directions of load strain applied) between carbon atoms C—C up to the fracture point, which is characterized by an

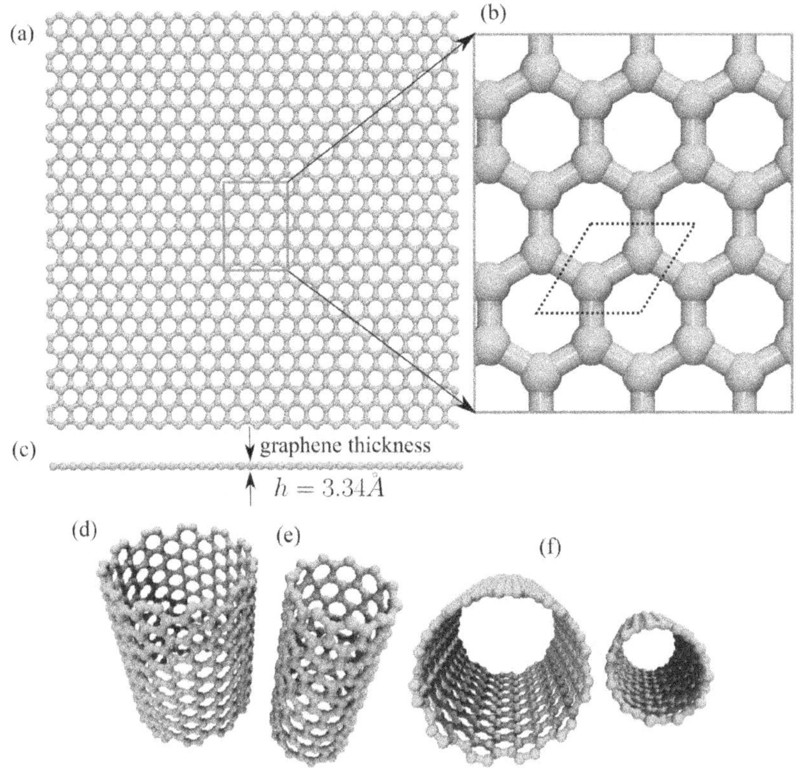

Figure 5.
Computational sample (graphene and CNT) in contact with a thermal reservoir R at a specific temperature different at 0.

abrupt drop of stress values to zero at 0.10 (0.13)—critical strain, respectively for X-direction (Y-direction). For higher temperatures (600 K and 1000 K), the stress/strain curves for graphene shows a reduction in ultimate tensile strength (UTS) and critical strain (σC) values. Those results for graphene were already observed for another theoretical investigation of graphene monolayer under thermal effects, see the bibliographic references [24, 25]. So, our results have good correspondence with results already obtained in the literature [26–28]. Therefore, the averages of the mechanical properties are listed in the following **Table 1**. In **Figures 6** and 7, we show the temporal atomistic evolutions of frames of the results obtained by reactive molecular dynamics simulations with an interatomic reactive force field ReaxFF, when the set of atomic configurations of load strain applied in graphene are in X and Y-direction and CNT.

In the results in **Figures 6** and 7, we can see the fully atomistic reactive molecular dynamics simulations with ReaxFF-potential for graphene monolayer for strain applied in X and Y-direction of mechanical load strain applied respectively. The results obtained whit ReaxFF potential show that the failure starts at C—C bonds which are aligned (in X and Y-direction of load strain applied). In all temperatures (300, 600, and 1000 K) we can see a clean failure rupture. The color bar in down of snapshot frames, shows the stress concentration in the monolayer the stretching dynamics, where the color blue are low-stress concentration and red color are high-stress concentrations in graphene monolayer. The results of CMD-ReaxFF for CNTs (armchair and zig-zag) (**Figure 7**), the stress is highly accumulated on the zigzag

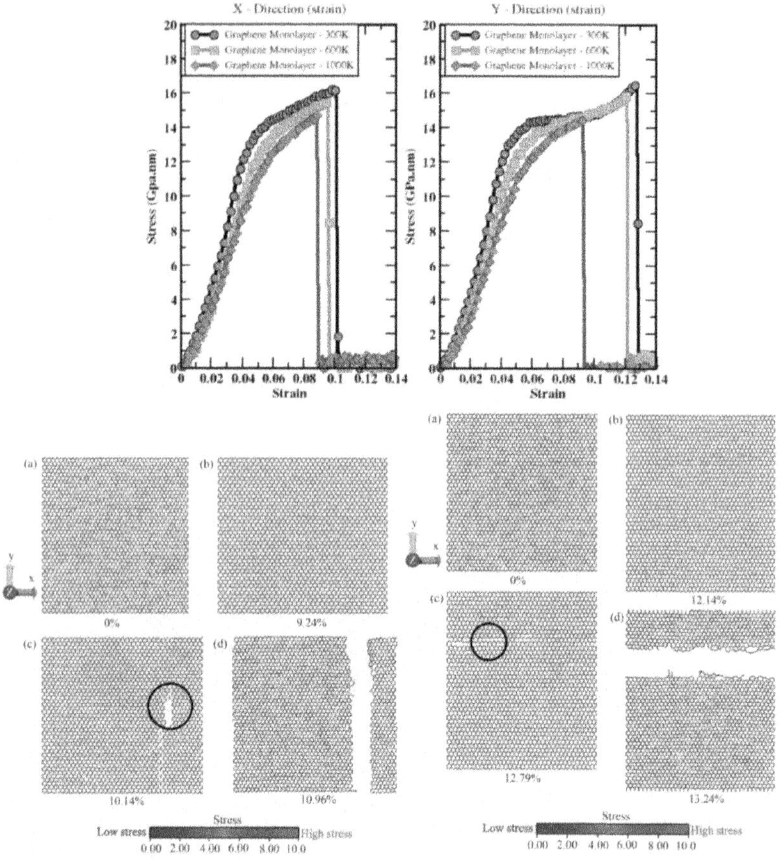

Figure 6.
Stress versus strain curves for graphene monolayers predicted by reactive molecular dynamics simulations with the ReaxFF interatomic potential at 300 K, 600 K and 1000 K in X-direction (left panels) and in Y-direction (right panels). Atomic frames representations of graphene monolayer under strain load in X-direction (room temperature): left side: (a) stretched (0%) of strain, in (b) 9.24% of strain, in (c) start break some chemical bonds C—C at 10.14% of strain and (d) the complete fracture of graphene monolayer at 10.96% of strain. In right side: atomic frames representations of graphene monolayer under load strain in Y-direction (room temperature): (a) un stretched (0%) of strain, in (b) 12.14% of strain, in (c) start break some chemical bonds C—C at 12.79% of strain and (d) the complete fracture of graphene monolayer at 13.24% of strain.

	Graphene monolayer			
	Temperature (K)	Y_{MOD}(**GPa nm**)	UTS(**GPa nm**)	**Critical strain**
X-Direction	300	3.14 ± 3.60	16.19 ± 0.03	0.10
	600	273.20 ± 3.17	15.63 ± 0.02	0.10
	1000	243.87 ± 2.27	14.76 ± 0.04	0.09
Y-Direction	300	3.13 ± 3.66	16.53 ± 0.02	0.13
	600	273.79 ± 3.37	15.86 ± 0.03	0.12
	1000	273.80 ± 2.88	15.86 ± 0.03	0.09

Table 1.
Mechanical properties values for graphene monolayer obtained by reactive molecular dynamics simulations with interatomic reactive potential ReaxFF calculated over a linear limit of 3%.

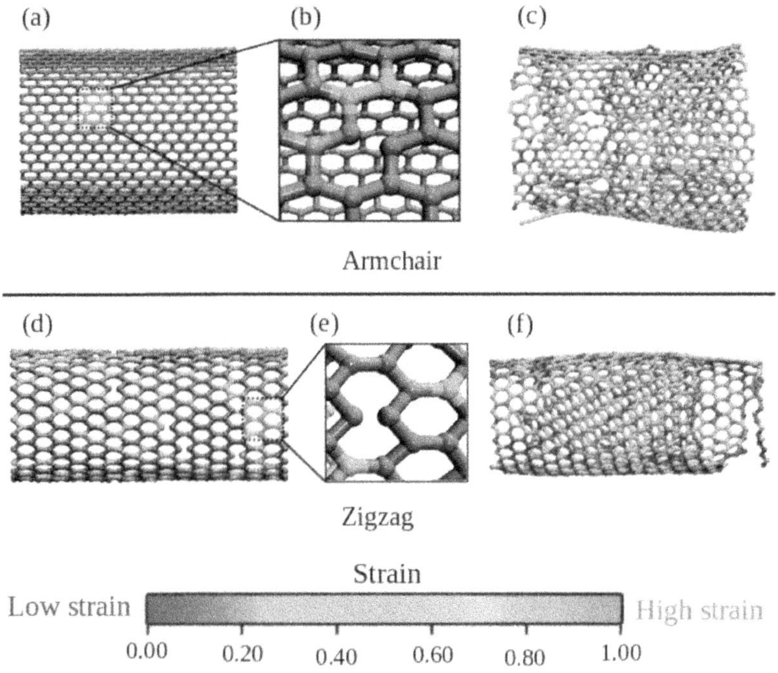

Figure 7.
Representative MD snapshots of a tensile stretch of conventional CNTs (armchair (11, 11) (top) and zigzag (11, 0) (bottom)). (a and d) Lateral view of the strained nanotube colored accordingly to the von Mises stress values (low stress in blue and high stress in red). (b and e) Zoomed view of the starting of bond breaking. (c and f) CMD-ReaxFF snapshot of the CNTs just after fracture [29].

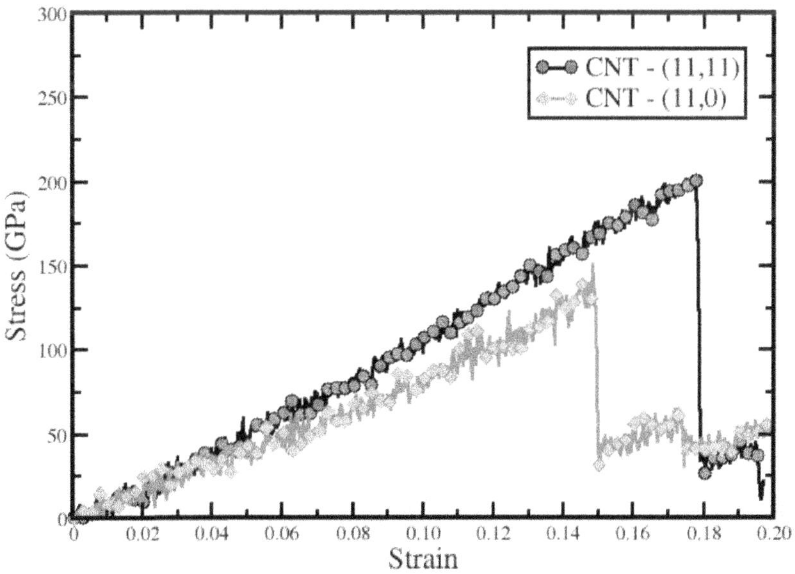

Figure 8.
Graphical representation of stress-strain curves obtained by CMD-ReaxFF for CNTs (11, 11)—black color and CNT (11, 0)—red color, at room temperature [29].

chains along the direction of the nanotube main axis. The fracture starts from the bonds parallel and nearly parallel to the nanotube main axis for the zigzag and armchair CNTs, respectively. Because CNTs lack the acetylene chains, the structure is more rigid, the stress is accumulated directly on the hexagonal rings, the critical strains are smaller, and the ultimate strength value is larger [29]. The obtained Young's modulus (see **Figure 8**) of the (11, 11) and (11, 0) CNTs were 955 GPa and 710 GPa, respectively. The ultimate tensile strength (UTS) 166 GPa and 122 GPa and $\sigma C = 18\%$ and $\sigma C = 16\%$ in good agreement with the average value of single-walled CNTs obtained by Krishnan et al. [30].

Author details

José Moreira de Sousa
Federal Institute of Education, Science, and Technology of Piauí—IFPI, São Raimundo Nonato, Piauí, Brazil

*Address all correspondence to: josemoreiradesousa@ifpi.edu.br

References

[1] Buehler Markus J, De Sousa JM. Atomistic Modeling of Materials Failure. Springer Science Business Media; 2018

[2] Allen MP, Tildesley DJ. Computer Simulation of Liquids. Oxford University Press; 2017

[3] Rapaport DC. The Art of Molecular Dynamics Simulation. Cambridge University Press; 2004

[4] Karplus M, McCammon JA. Molecular dynamics simulations of biomolecules. Nature Structural Biology. 2002;**9**(9):646-652

[5] De Sousa JM. Dinâmica molecular reativa de sistemas nanoestruturados [tese]. Repositorio da Producao Cientifica e Intelectual da Unicam Produção Científica Instituto de Física "Gleb Wataghin"—IFGW; 2016

[6] Gates TS, Odegard GM, Frankland SJV, Clancy TC. Computational materials: Multi-scale modeling and simulation of nanostructured materials. Composites Science and Technology. 2005;**65** (15-16):2416-2434

[7] Liu WK, Karpov EG, Zhang S, Park HS. An introduction to computational nanomechanics and materials. Computer Methods in Applied Mechanics and Engineering. 2004;**193**(17-20):1529-1578

[8] Boldon L, Laliberte F, Liu L. Review of the fundamental theories behind small angle X-ray scattering, molecular dynamics simulations, and relevant integrated application. Nano Reviews. 2015;**6**(1):25661

[9] Novoselov KS, Geim AK, Morozov SV, Jiang DE, Zhang Y, Dubonos SV, et al. Electric field effect in atomically thin carbon films. Science. 2004;**306**(5696):666-669

[10] Iijima S. Helical microtubules of graphitic carbon. Science Nature. 1991; **354**(6348):56-58

[11] Goldstein H, Poole C, Safko J. Classical Mechanics. 2002

[12] Landau LD, Lifshitz EM. Course of Theoretical Physics. Elsevier; 2013

[13] Marion JB. Classical Dynamics of Particles and Systems. Academic Press; 2013

[14] Martys NS, Mountain RD. Velocity Verlet algorithm for dissipative particle-dynamics-based models of suspensions. Physical Review E Nature. 1999;**59**(3): 3733

[15] Rappé AK, Casewit CJ, Colwell KS, Goddard WA III, Skiff WM. UFF, a full periodic table force field for molecular mechanics and molecular dynamics simulations. Journal of the American Chemical Society. 1992;**114**(25): 10024-10035

[16] Mueller JE, Van Duin AC, Goddard WA III. Development and validation of ReaxFF reactive force field for hydrocarbon chemistry catalyzed by nickel. The Journal of Physical Chemistry C. 2010;**114**(11):4939-4949

[17] Van Duin AC, Dasgupta S, Lorant F, Goddard WA. ReaxFF: A reactive force field for hydrocarbons. The Journal of Physical Chemistry A. 2001;**105**(41): 9396-9409

[18] Su JT, Goddard WA III. Excited electron dynamics modeling of warm dense matter. Physical Review Letters. 2007;**99**(18):185003

[19] Senftle TP, Hong S, Islam MM, Kylasa SB, Zheng Y, Shin YK, et al. The ReaxFF reactive force-field: Development, applications and future

directions. npj Computational Materials. 2016;**2**(1):1-14

[20] Plimpton S. Fast parallel algorithms for short-range molecular dynamics. Journal of Computational Physics. 1995; **117**(1):1-19

[21] Clarke L, Glendinning I, Hempel R. The MPI message passing interface standard. In: Programming Environments for Massively Parallel Distributed Systems. Basel: Birkhäuser; 1994. pp. 213-218

[22] Evans DJ, Holian BL. The Nose–Hoover thermostat. The Journal of Chemical Physics. 1985;**83**(8): 4069-4074

[23] Salinas S. Introduction to Statistical Physics. Springer Science Business Media; 2001

[24] Goyal M, Gupta BRK. Study of shape, size and temperature-dependent elastic properties of nanomaterials. Modern Physics Letters B. 2019;**33**(26): 1950310

[25] Shen L, Shen HS, Zhang CL. Temperature-dependent elastic properties of single layer graphene sheets. Materials and Design. 2010; **31**(9):4445-4449

[26] Brandão WHS, Aguiar AL, De Sousa JM. Atomistic computational modeling of temperature effects in fracture toughness and degradation of penta-graphene monolayer. Chemical Physics Letters. 2021:138793

[27] Cadelano E, Palla PL, Giordano S, Colombo L. Nonlinear elasticity of monolayer graphene. Physical Review Letters. 2009;**102**(23):235502

[28] Lee C, Wei X, Kysar JW, Hone J. Measurement of the elastic properties and intrinsic strength of monolayer graphene. Science. 2008;**321**(5887): 385-388

[29] De Sousa JM, Bizao RA, Sousa Filho VP, Aguiar AL, Coluci VR, Pugno NM, et al. Elastic properties of graphyne-based nanotubes. Computational Materials Science. 2019; **170**:109153

[30] Krishnan A, Dujardin E, Ebbesen TW, Yianilos PN, Treacy MMJ. Young's modulus of single-walled nanotubes. Physical Review B. 1998; **58**(20):14013

Chapter 3

Elements of the Nonlinear Theory of Elasticity Based on Tensor-Nonlinear Equations

Kirill F. Komkov

Abstract

The chapter contains information that forms the basis of a new direction in the nonlinear theory of elasticity. The theory, having adopted the mathematical apparatus obtained in the middle of the last century, after its analysis, is used with significant changes. This concept allows us to more accurately reveal the mechanism of deformation of materials, the elastic nature of which significantly depends on the type of stress state, due to the growth of additional volumetric deformation associated with the accumulation of defects, called dilatation. The work is original — after abandoning the elasticity characteristics in the form of modules - constants, the main role is assigned to material functions, which represent statistical characteristics. Their relation can be considered a coefficient of variation and a parameter of tensor nonlinearity, which makes it possible to represent the deformation in the form of two parts, different in origin.

Keywords: dilatancy, volume deformation, shape change, phase similarity of deviators, volume deformation, coefficient of variation, tensor nonlinearity, anisotropy, variable elasticity parameter

1. Introduction

Experimental studies of well-known mechanics with various materials already in the eighteenth century revealed numerous nonlinear effects described in the book [1]. From the standpoint of the linear theory of elasticity, many of them could not be explained, so they were called second-order effects, as not significant. However, in the middle of the twentieth century, they pushed M. Rayner [2], and a little later, V. V. Novozhilov [3], to the need to develop a theory based on a new concept of tensor-nonlinear equations [4, 5] that more accurately reflect the nonlinearity of materials. The widespread introduction of composite media and the study of their mechanical properties began at the end of the last century. In the same years, a lot of experimental works appeared to study the mechanical properties of various composites, illuminating the properties of not only reinforced materials, but also grain composites, which differ in different reactions to tension and compression. This property is possessed by media whose longitudinal modulus of elasticity and other characteristics depend on the type of stress state, determined at values of deformations close to zero. It should be called the work of Tolokonnikov L. A., Makarov E. S. [6] and many others who have devoted research to the properties of these media, in which the presence of damage to internal connections and loosening, that is, the development of dilatancy, is stated. The theories put forward by them are based on tensor-linear equations. As a rule, in them all the characteristics of different-modulus media are determined from the condition of the existence of a specific deformation potential.

In this paper, in continuation of the study [7], to take into account the noted effects, such a transformation equations was found, which made it possible to develop methods for determining the elasticity characteristics. These equations presented for the main deformations made it possible not only to describe the deformation of the shape change, the coefficients of transverse deformations along different axes, to determine the volume deformation depending on the average stress, but also the dilatancy associated with the shape change.

2. About of different-module materials

The development of methods was carried out based on the results of studies of grain composite [8], and in earlier works of gray cast iron, using the research of [9]. The first is a hardened mechanical mixture of a mineral filler with a polymer matrix, the test results and information about its properties are published in [8, 10–12]. These materials have not only the presence of divergence of the initial longitudinal modules under tension and compression, but also show the dependence of elastic properties on time; therefore, in this work, the test results obtained at a single strain rate are used. The nonlinearity of the diagrams of a grain composite is clearly represented by the results of testing cross-shaped samples under repeated static stretching. It has a high malleability at normal temperature. The main purpose of testing such samples was to more fully reveal the mechanism of deformation of different-modulus materials. **Figure 1a** shows the curve 1—the ascending branch at the first cycle of active deformation along the axis 1–1 represents the initial properties of the material. Where P is the force in H, Δl is the elongation in millimeters. When unloading, the curve decreases sharply, which indicates a significant decrease in the number of bonds that break down with small deformations. The residual deformation does not represent plastic properties, but a residual dilatancy, from which it is possible to make a quantitative assessment of the initial deformation anisotropy for the next loading cycle. Curve 2—the ascending branch of the second cycle illustrates the resistance of the restored "short" and remaining "long" bonds. In **Figure 1b**, curve 1 is the ascending branch of the test at the first cycle along the axis 2–2. For comparison, a diagram (dashed) is shown, marking the initial properties of the composite. The difference in the curves of the first cycles in different directions suggests that the connection break occurs in the transverse direction as well. The first curve shows that the "short" connections in the direction 2–2 are partially preserved.

The difference between the ascending branches of the first and second stretching cycles along the 1–1 and 2–2 axes is a real one, called [3] by V. V. Novozhilov "real" anisotropy. The second cycle shows that the material has noticeably softened, the slope of the curve has decreased, but the tangential longitudinal elastic modules manifest themselves on the second part of the branch as increasing, differing from the first cycle. This emphasizes the fact that the links are divided into "short" and "long"—stronger, although in [13] a more detailed gradation of links is given, which will be superfluous for this work.

Both in [8, 12], it is noted that stretching is accompanied by a noticeable increase in volume. The same is observed with compression, although to a lesser extent. The loss of bonds and softening are the cause of the loss of elastic energy, which is taken

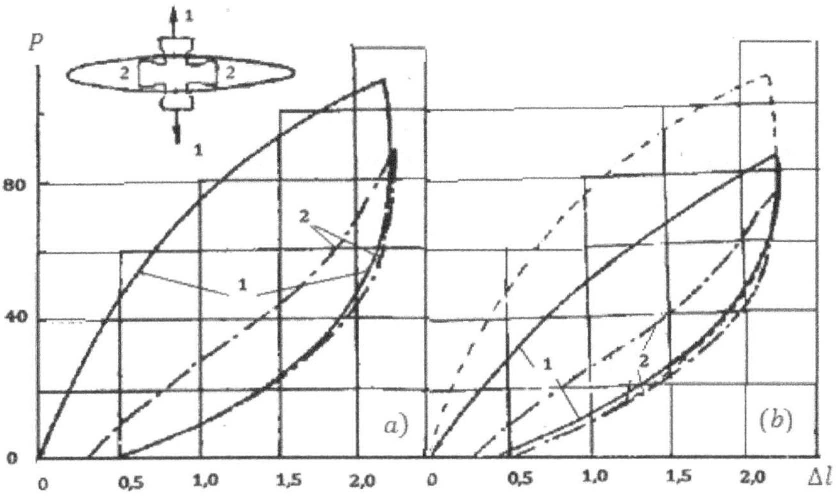

Figure 1.
a—Curve 1—The ascending branch at the first cycle of active deformation on the axis 1–1, curve 2—The ascending branch of the second cycle; b—Curve 1—The ascending branch at the first cycle on the axis 2–2.

into account by the mathematical model with a proportional increase in stresses only by the growth of additional volume deformation, as in the deformation theory, plastic shifts. For practical calculations, test diagrams of standard samples were used according to the method described in [8]. The tensile diagram for testing along the 1–1 axis, curve 1, **Figure 1a**, is a sequence of limit values of groups of bonds that are close in strength. The same is true for other types of loading, but to a lesser extent.

The purpose of this work is to fully reveal the possibilities tensor-nonlinear equations: transformed to a form convenient for the formulation of material functions, analysis, and processing of test results. On their basis, to develop methods for calculating all characteristics, including the coefficients of transverse deformations, elastic modulus, and compliance, as well as parameters that characterize the loosening of the structure and the change in elastic properties both with increasing load and with a change in the type of stress state.

3. On tensor-nonlinear equations

To describe the deformation of different-modulus materials, considering them isotropic, we used tensor-nonlinear equations of the connection of the strain deviator D_e with the stress deviator D_σ by V. V. Novozhilov [3], which, unlike the equations of M. Reiner [2], do not yet require the equation of the connection of the average strain with the average stress:

$$\varepsilon_{ij} - \frac{1}{3}\hat{e}_1\delta_{ij} = \frac{1}{2G}\left[\frac{\cos(2\xi+\psi)}{\cos 3\xi}S_{ij} + \sqrt{\frac{3}{\hat{s}_2}}\frac{\sin\omega}{\cos 3\xi}\left(S_{i\alpha}S_{\alpha j} - \frac{2}{3}\hat{s}_2\delta_{ij}\right)\right]. \tag{1}$$

In the left part: $e_{ij} = \varepsilon_{ij} - \varepsilon_0\delta_{ij}-$ components of the strain deviator; $\varepsilon_0 = (\varepsilon_{ii})/3 = \hat{e}_1/3-$ average strain; $\hat{e}_1$ – the first, $\hat{e}_2 = 3e_0^2/4$ – the second and $\hat{e}_3 = 3\det|D_e|-$ the third invariants of the strain tensor;

$$e_0 = \left(2/3e_{ij}e_{ij}\right)^{1/2} \tag{2}$$

Strain intensity. In the right part: $S_{ij} = \sigma_{ij} - \sigma_0\delta_{ij}$– components of the stress deviator; $\sigma_0 = (\sigma_{ii})/3 = \hat{s}_1/3$– medium voltage, $\hat{s}_1$– first, $\hat{s}_2 = S_0^2/3$– second and $\hat{s}_3 = -3\det|D_\sigma|$–third invariants of the stress tensor; $S_0 = \left(32S_{ij}S_{ij}\right)^{1/2}$– is the intensity of the stress; $S_i = S_0c_i/3$- principal values of the stress deviator; $e_i = e_0d_i/2$– the main values of the deviator of the strain used in [3]; $c_1 = 2\cos\xi$, $c_2 = \sqrt{3}\sin\xi - \cos\xi$, $c_3 = -\left(\sqrt{3}\sin\xi + \cos\xi\right)$– trigonometric values that relate the main stresses to the stress intensities and similar d_i to the strain intensities.

Abandoning the constancy of the phase similarity diverters ω, which was proposed in [4], the generalized modulus G and the phase can be expressed through the coefficients of the tensor arguments:

$$X = \frac{1}{2G}\frac{\cos(2\xi+\psi)}{\cos 3\xi}, \quad Y = \frac{1}{2G}\sqrt{\frac{3}{\hat{s}_2}}\frac{\sin\omega}{\cos 3\xi} = \frac{1}{2G}\frac{3}{S_0}\frac{\sin\omega}{\cos 3\xi} \tag{3}$$

For this we can use Eq. (1) presented for the main component of the deviator of the strain

$$e_i = XS_i + Y\left(S_i^2 - 2/9S_0^2\right). \tag{4}$$

The coefficients X and Y can be given an unambiguous physical meaning and formulas for determining them can be derived. Using three shear pliabilities $\varphi_i = \gamma_i/\tau_i$ i in sites with principal tangential stresses $\tau_i = (S_j - S_\alpha)/2$, where $\gamma_i = e_j - e_\alpha$– a are the principal shifts, Eq. (12) allow us to find three shear pliabilities $\varphi_i = 2\left(X - \hat{Y}c_i\right)$, where $\hat{Y} = YS_0/3$. Given that the sum $(c_i) = 0$, from the relations for the pliabilities we find their average value and standard deviation:

$$\Phi_m = (\phi_i)/3 = 2X; \quad \Phi_d = \left\{\left[\left(\phi_j - \phi_\alpha\right)^2\right]/8\right\}^{1/2} \tag{5}$$

Thus, the analysis of the Eq. (1) allows, without any assumptions, to be free from uncertainty and to find an approach to the characterization of the deformation Φ_m and Φ_d that are already used for different materials, therefore will continue to remain the same notation, naming the material features:

$$\Phi_m = 2X = \frac{\cos(2\xi+\psi)}{G\cos 3\xi};. \quad \Phi_d = \frac{1}{3}\hat{Y} = \frac{\sin\omega}{2G\cos 3\xi} \tag{6}$$

The sum of the squares of the differences of the main values of the deformation deviator

$$\left(e_i - e_j\right)^2 = \frac{S_0{}^2}{9}\left(c_i - c_j\right)^2\left(X^2 + 2X\hat{Y}c_\alpha + \hat{Y}^2c_\alpha^2\right) \tag{7}$$

leads to the need to calculate the relations: $\sum\left(c_i - c_j\right)^2 = 18$, $\sum c_\alpha\left(c_i - c_j\right)^2 = 18\sin 3\xi$, $\sum c_\alpha^2\left(c_i - c_j\right)^2 = 18$; $i, j, \alpha = 1, 2, 3; i \neq j \neq \alpha$. Finally, the relationship between the strain intensity (2) and the stress intensity is reduced to the equation:

$$e_0 = \frac{2S_0}{3}\left[\left(X^2 + 2X\hat{Y}\sin 3\xi + \hat{Y}^2\right)\right]^{1/2}. \tag{8}$$

It leads to generalized malleability:

$$\Phi_\xi = \frac{3e_0}{S_0} = \left[\Phi_m^2 + (4/3)\Phi_m\Phi_d\sin 3\xi + (4/9)\Phi_d^2\right]^{1/2}, \tag{9}$$

as a function of the angle ξ, and the inverse of the malleability to the generalized modulus of elasticity under shear:

$$G = 1/\Phi_\xi = \frac{1}{2}\sqrt{\hat{s}_2/\hat{e}_2} = 1/\left\{2\left[\left(X^2 + 2X\hat{Y}\sin 3\xi + \hat{Y}^2\right)\right]^{1/2}\right\}. \tag{10}$$

It follows from this relation that the modulus clearly depends on the type of stress state, and it can be a constant value only in the special case, as it was envisaged in [4]. After replacing the second invariants on the stress intensity and strain intensity and replacing the sequence of main stresses: $\sigma_1 \geq \sigma_2 \geq \sigma_3$, it is possible to give the original (1) equations of V. V. Novozhilov a form that was used without simplifications in the work [7]. After replacing the second invariants on the stress intensity and strain intensity and replacing the sequence of main stresses:, it is possible to give the original equations of V. V. Novozhilov (1) the form, without any simplifications, which was also used in the work [7]:

$$e_{ij} = \Phi_m S_{ij}/2 + \Phi_d\left(S_{i\alpha}S_{\alpha j} - 2S_0^2/9\delta_{ij}\right)/S_0. \tag{11}$$

After replacing the third invariants, the formulas for the angles take the form: the first $\xi = 1/3\arccos\left[27S_{ij}S_{j\alpha}S_{\alpha i}/\left(2S_0^3\right)\right]$ – is the angle of the stress state view, and the second one is $\psi = 1/3\arccos\left[4e_{ij}e_{j\alpha}e_{\alpha i}/\left(3e_0^3\right)\right]$ – the angle of the view of the deformed state, which change already in other limits: $0 \leq \xi\ u\ \psi \leq \pi/3$; $i, j, \alpha = 1, 2, 3$; $i \neq j \neq \alpha$. The coefficients for tensor arguments (6) make it possible to find a formula for determining the phase the similarity of deviators:

$$\omega = \xi - \psi. \tag{12}$$

The exact definition of which is given below. Performing trigonometric transformations taking into account the new sequence of principal stresses, the material functions in Eq. (8) can be represented:

$$\Phi_m = \Phi_\xi \sin(3\xi - \omega)/\sin 3\xi = \varphi_i/3; \tag{13}$$

$$\Phi_d = 3\Phi_\xi \sin(\omega)/(2\sin 3\xi) = \left\{3/8\left[(\Phi_m - \varphi_i)^2\right]\right\}^{1/2}. \tag{14}$$

where they acquire values that have a physical meaning of average and standard compliance, manifesting themselves by statistical characteristics. The deviatory part of M. Rayner's equations [2] leads to the same results of the functions φ_i.

4. Initial data

Due to the lack of proven methods, the first calculations in [8] used only the results of tensile and compression tests. Generalized compliance is determined by the relation (6), which for these states is taken by simple expressions:

$$\Phi_{\xi p} = \Phi_m + 2/3\Phi_d, \quad \Phi_{\xi c} = \Phi_m - 2/3\Phi_d \tag{15}$$

Assuming the independence of these functions from the type of stress state, we find a simple way to approximate the calculation of the shear modulus and the phase similarity of deviators according to the formula (6). The form change for any stress state, although approximate, can be described. To refine it, you can use the same ratio, but for a pure shift. At the same time, difficulties arose due to the fact that the tests were usually carried out on other equipment and other means of measuring deformations, so the lack of initial data was compensated by algorithms that were derived from the same equations converted to equations for anisotropic media [8, 10].

Experimental data obtained by tensile testing and compression of grain composite [8, 11], which has the maximum deformation under compression $\varepsilon_c > 10\%$ in the form of primary charts $\sigma_i - \varepsilon$ and graphs for the coefficients of lateral deformations, $\nu_i - \varepsilon$, $\nu_i = -\varepsilon_n/\varepsilon$, stress and strain have to specify, according to the formulas of [4], which can be given as follows:

$$\sigma^* = \sigma\left[\left(1 + 2\varepsilon_n^*\right)^2/\left(1 + 2\varepsilon^*\right)\right]^{1/2}, \tag{16}$$

where $\varepsilon^* = \varepsilon(1 + \varepsilon/2)$; $\varepsilon_n^* = \varepsilon_n(1 + \varepsilon_n/2)$; $\nu_i' = -\varepsilon_n^*/\varepsilon^*$ – are the coefficients of transverse deformations; $i = p, c$; index n – indicates transverse deformation. The stresses σ^* and deformations ε^* are called reduced [4]. Then the asterisk above the given stresses and deformations is removed. The ratio of material functions can be considered a coefficient of variation [14, p. 544]:

$$p = \Phi_d/\Phi_m = 3\sin\omega/2\sin(3\xi - \omega), \tag{17}$$

Since the material functions exhibit a statistical character, and its values correspond to the condition: $p < 1$. The study of its extremum shows that the derivative with respect to the angle ξ is zero if the phase of similarity of deviators obeys equality:

$$\omega = \mathrm{arctg}[2p\sin3\xi/(3 + 2p\cos3\xi)]. \tag{18}$$

The graphs for the phase differ slightly from the half-wave of the sine wave when the angle ξ changes from zero to $\pi/3$.

For phase values other than zero, the ratios of the deviator components belonging to the same stress state are not equal: $e_1/S_1 \neq e_2/S_2 \neq e_3/S_3$– is the condition of their disproportionality. However, for the states of tension and compression, this inequality becomes an equality: $S_1/S_2 = e_1/e_2 = 1$, since similarity conditions are implemented for them, since $S_2 = S_3$ and $e_2 = e_3$, so the phase is zero regardless of the properties of the This conclusion is consistent with the relation (13), which directly follows from the formulas (13) and (10). The material functions are similar: $\Phi_d = p\Phi_m$– for all states. This connection of functions allows us to consider both shape-changing deformations and volumetric deformations in the form of two parts. The first part should be associated with a change in the intermolecular distances in the rigid elements of the structure, and the second part of the deformation, including the coefficient of variation, should be attributed to the loss of bonds [8]. These deformations, despite their different physical origin, are included in the model as elastic. The initial data were taken based on the results of tests [8] obtained during tests of grain composites, the diagrams of which are shown in **Figure 2a** with dashed lines.

Solid lines represent two diagrams, after the refinement performed according to formulas (11). The dependence of $S_{0\tau}$ on the strain, taken as a diagram for pure shear (fine stroke), is obtained from diagrams for stretching and compression, according to an algorithm [7] using transformed equations. The stress values along the ordinate axis in **Figure 2a** in MPa.

Graphs for the coefficient of variation p (dashed line), the maximum values of the similarity phase of the deviators ω_{max} (small stroke), and the functions by which they are determined are shown in **Figure 2b**. These functions include Φ_d and Φ_ξ for stretching and compressing (solid lines).

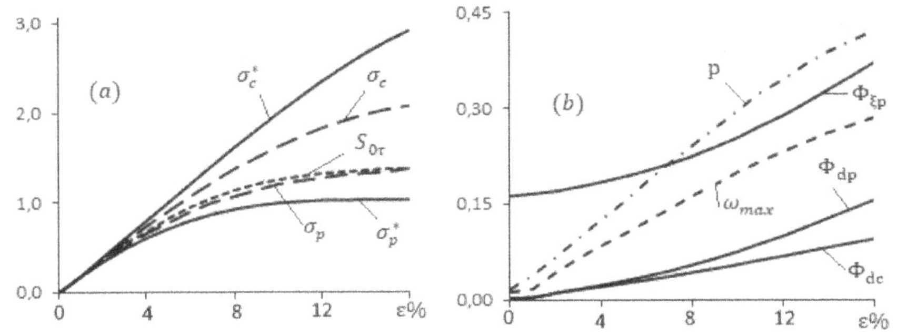

Figure 2.
a: Test diagrams of granular composites: Curve σ_p– During the tensile test, curve σ_c– for compression, curve $S_{0\tau}$–according to the algorithm using data on tension and compression; curves σ_p^* and σ_c^* – after the transition to the reduced stresses. b: Curves based on the results of calculations: The change in the p – Coefficient of variation and the ω–phase of the similarity of deviators and the curves Φ_d, Φ_m, and Φ_ξ for the characteristics of the shape change with increasing deformation.

5. On the equations for form-changing

The rejection of the constancy of the phase gives the ratio of (6), which after the transition to the second sequence of the principal stresses is the law of deformation:

$$e_0 = \Phi_\xi S_0/3, \tag{19}$$

where the main characteristic becomes generalized compliance (7):

$$\Phi_\xi = \Phi_m\left[1 + (4/3)p\cos 3\xi + (4/9)p^2\right]^{1/2} = 1/G, \tag{20}$$

as the inverse of the generalized shift modulus of G, they are represented in a discrete (digital) form by a mathematical model, as well as material functions. After replacing the sequence of main stresses, sin3ξ in the expression (6) is transformed in the ratio (15) into $\cos 3\xi$. If the relation (15) is simplified by getting rid of the square root, then the second part with the coefficient of variation can be represented as:

$$e_0^* = \Phi_\xi^* S_0/3, \quad \Phi_\xi^* \approx p\Phi_m[\cos 3\xi + (1/3)p] \tag{21}$$

where the compliance for the second part is the value Φ_ξ^*.

From the ratio (15) for stretching and compression, it also follows:

$$\Phi_{\xi i} = \Phi_{mi}(1 \pm 2/3p), \tag{22}$$

where $i = p, c$; (p– stretching, c - compression). The functions Φ_{mi} and Φ_{di}, as the characteristics of the shape change, are determined for these states using the first Cauchy sign [14]. On this basis, their values follow from the relations (13) and (10), if the angle ξ is shifted by a small deviation from the original angles. The second variant of determining the coefficient of variation follows from the relations (16):

$$p = 3(\kappa - \kappa_m)/2(\kappa + \kappa_m). \tag{23}$$

It protects the characteristics of the shape change from errors in their calculations: Φ_m, Φ_d and Φ_ξ, where $\kappa = \Phi_{\xi p}/\Phi_{\xi c}$– is the ratio of generalized and $\kappa_m = \Phi_{mp}/\Phi_{mc}$– is the average compliance. Calculation of material functions by formulas (13) and (10), or rather by their second equalities, cannot be carried out, since there is no initial information about the functions $\varphi_i = \gamma_i/\tau_i$ for the same state. This obstacle can be overcome if we use the following postulates: the first one states that the values of the functions φ_i can be considered the values of the malleability $\varphi_i = 3e_0/S_0 = \Phi_{\xi i}$ for three stress states: stretching, net shear, and compression. According to the second one, the functions $\varphi_i = \Phi_{\xi i}$are equal.

The results of calculations for two variants according to the formulas (12) and (17) showed that they differ only by the fifth significant digit after the decimal point for any loading stage. It is for checking the postulates that duplication is necessary. If there is a coefficient of variation, the calculation of material functions for any other states is significantly simplified: first, Φ_m by relation (13) is determined, and then $\Phi_d = p\Phi_m$, as a function of the angle ξ and the load level, since the coefficient of variation is the only value independent of the type of stress state.

6. On the equation for volumetric deformation

The derivation of equality (21), as an additional part of the deformation of the form change, is proposed as an unknown formula for dilatancy, as a part of the volume deformation, consistent with the previously expressed idea that the parameter p allows the deformation, divided into two parts. This thought, the results of experimental studies and already published works allow us to propose an equation for the volumetric strain in the following form:

$$\varepsilon_0 = \varepsilon_y + \varepsilon_g = \sigma_0/3K_\xi + 2p\Phi_m æ S_0(1 + k\zeta)/9. \tag{24}$$

The first part ε_y– linearly dependent on the mean stress refers to the deformation of the stiffer elements of the structure, where the value K_ξ is the theoretical bulk elasticity modulus. The formula for linear-elastic deformation is inherited from linear elasticity theory, and the second part ε_g– dilatancy with the parameter p, including œ – the loosening parameter and Φ_d– the function reflecting the dependence of the volume strain on the form change. The coefficient k in formula (18) was introduced in order to take into account the influence of average stress on dilatancy as well as for convenience of checking the proposed relation. So, at k = 0 the formula for dilatancy takes the form that has already been used in several works of the author, including [7, 15], because at k = 0.3 the curves for volume deformations under tension and compression are well superposed on the experimental curves.

The process of transformation of the tensor-nonlinear equations mentioned above is covered in sufficient detail in [7 , p. 56] and probably first implemented in [10]. The equations for coupling the strain tensor to the stress tensor (8), together with the equation for average strain with average stress (18), lead to the equations for coupling the strain tensor to the stress tensor

$$\varepsilon_{ij} = \frac{(3\Phi|k + 2œ\Phi_d k)\sigma_0\delta_{ij}}{9} + \frac{\Phi_m S_{ij}}{2} + \frac{\Phi_d}{S_0}\left[S_{i\alpha}S_{\alpha j} - \frac{2(1-œ)S_0^2\delta_{ij}}{9}\right]. \tag{25}$$

The equations reduced to the principal deformations are used for the matrix transformation: $\varepsilon_i = a_{ij}\sigma_i$, which can then be reduced to the form of equations characteristic of anisotropic media:

$$\varepsilon_i = \sigma_i/E_i - \nu_{ji}\sigma_j/E_j - \nu_{\alpha i}\sigma_\alpha/E_\alpha, \tag{26}$$

with the known specifications for the diagonal components:

$$a_{ii} = [3\Phi_m + \phi_h + \Phi_d c_{ii}] = E_i^{-1} \tag{27}$$

and non-diagonal matrix components:

$$a_{ij} = \left[\frac{3\Phi_m}{2} - \phi_h - \Phi_d c_{ij}\right] = -\nu_{ij} E_j^{-1}, \tag{28}$$

where $\phi_h = 1/K_\xi + 2\Phi_d æk/9 = (\phi_\xi + a\Phi_d)/3$; $a = 2kœ/3$; E_i–moduli of longitudinal elasticity, ν_{ij}– coefficients of transverse deformation.
Reconciliation of Eqs. (7.6) leads to the equation for the relation of average strain with stresses:

$$\varepsilon_0 = (\sigma_1 \phi_{k1} + \sigma_2 \phi_{k2} + \sigma_3 \phi_{k3})/3, \tag{29}$$

where $\phi_{ki} = 1/K_i$ is the bulk elasticity yield

$$\phi_{ki} = 3(1 - \nu_{ij} - \nu_{i\alpha})/E_i = 3(1 - \nu_i)/E_i. \tag{30}$$

Pairs of coefficients $\nu_i = (\nu_{ij} + \nu_{i\alpha})/2$ determine the transverse deformations in three directions of the main stresses and volumetric deformations; where $i = 1, 2, 3; i \neq j \neq \alpha$. The relations (23) are an integral part of the methodology of determining K_ξ– theoretical bulk modulus of elasticity and œ– the loosening parameter. In this process, the most critical importance is assigned to the procedure of matching theoretical curves for transverse strain coefficients [7].

7. Supplement to the methodology

The high values of the theoretical modulus of volumetric elasticity, but low for compliance with tension, and low for compression, can be explained by a simple transformation of the ratio (18), if we isolate from it $\varepsilon_y = \varepsilon_{0i} - \varepsilon_{gi} = \sigma_0 \phi_{ki}/3$–linear volumetric deformation. It allows you to find the pliability ϕ_{ki} for stretching and compression, which are required to combine experimental curves with theoretical curves during the transformation. Taking $\zeta_c = -\zeta_p$, $1/\zeta_i \cong \pm 3$; $1/\zeta = 3,0009$ и $K_i = E_i/3(1 - 2\nu_i)$, simple actions lead to the formulas:

$$\phi_{kp} = \frac{1}{K_p} - 2œ_p \Phi_{dp}(1/\zeta + k) \cong \frac{1}{K_p} - 6.6œ_p \Phi_{dp}, \tag{31}$$

$$\phi_{kc} = \frac{1}{K_c} + 2œ_c \Phi_{dc}(1/\zeta - k) \cong \frac{1}{K_c} + 5.4œ_c \Phi_{dc}. \tag{32}$$

It follows from the first that the second term reduces the flexibility for stretching, and the value of the theoretical module, on the contrary, increases as an inverse value. In the second formula, the second term increases the malleability for compression, although dilatancy is present. The second terms in these relations allow us to quantify its influence on the values of theoretical compliance. From the second formula, for compression, greater malleability is required, although dilatancy is present. The second terms in these relations allow us to quantify its influence on the values of theoretical compliance. Since the pliability of ϕ_{kp} is determined by the initial value of the function Φ_{dp}, it makes sense to refine it by redefining the loosening parameter $œ_p \cong \left(\phi_p - \phi_{kp}\right)/6.6\Phi_{dp}$ and then dilatancy. The mean stress in pure shear is zero, but given that, $\zeta_p + \zeta_c = 0$, as the value of the parameter ζ_τ, it is suggested algorithm, as a response to the question about the significance of the theoretical module, and for this condition: $\phi_{k\tau} = \frac{1}{K_\tau} + œ_\tau \Phi_{d\tau} \approx K_{\xi i}/2$, where $i = p, c$; (p– stretching, c– compression).

It follows from the relations (24) and (25) that in the process of converting tensor-nonlinear equations to matrix equations, the pliabilities $\phi_{ki} = 1/K_{\xi i}$ are realized, the values of which along the axes 2 and 3 satisfy the conditions of continuity and smoothness, as functions of the main stresses. These formulas contain an answer to the reasons for the large difference in the values of the theoretical module. It is called theoretical, because its values correspond to the inequality with respect to the classical module: $K_\xi > K$. The considered technique made it possible to find such values of the theoretical elastic modulus that lead to more accurate values of the linear elastic volume deformation.

8. On deformation anisotropy

V. V. Novozhilov in his work [3] expressed his opinion about this phenomenon, for the description of which the mathematical apparatus of tensor-nonlinear equations can be used, as an "important phenomenon," without emphasizing on what characteristics it manifests itself. The studies show that the effect of dilatancy on the longitudinal elastic moduli E_i is not significant. Their divergence with different indices is less than 5%, but leads to appreciable strain anisotropy of the transverse strain coefficients. In the history of the mechanics of materials described in the book [1], much space is devoted to the research of its initial value (the Poisson's ratio), since not only modules, but also the theories of authoritative scientists depended on it. However, the latter values, for example, at destructive stresses, are not given due attention, especially in other areas of the main stresses. In this paper, perhaps for the first time, graphs of the theoretical coefficients of transverse deformations are given. They are easier to describe not by formulas, but by graphs for: $\nu_{12}, \nu_{31}, \nu_p, \nu_c$, and ν_i, $\Sigma\nu_i/3$. The line in **Figure 3a**, represented by points, here repeats the curves for $\nu_{12} = \nu_{13}$, which are combined with the values of the coefficient ν_p by the method. The deviation of the curve for the coefficient ν_p from its initial value should be considered the main "source" of dilatancy and all other coefficients. If this curve for the coefficients ν_p and $\nu_1$coincided with the graph for $\Sigma\nu_i/3$, then all the curves presented in

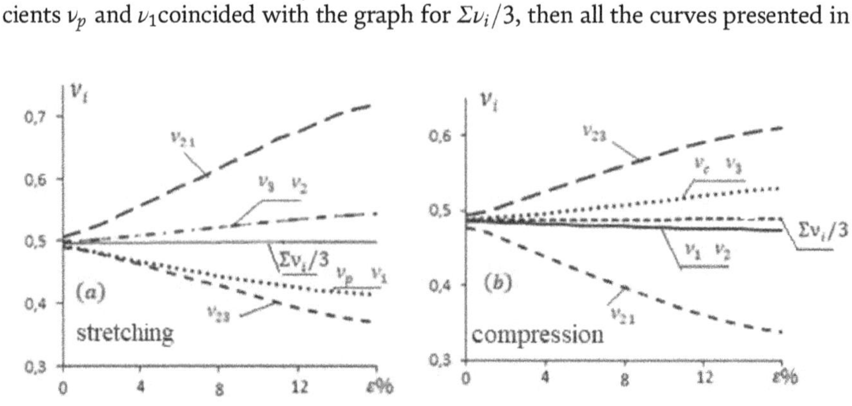

Figure 3.
The change in the coefficients of transverse deformations: a – Under tension; b – Under compression.

Figure 3a, would merge into one curve, and there would be no dilatancy. The main direction is the voltage σ_1.

The lower the values of the last points of the curve for ν_p fall, the greater the dilatancy takes on and the higher the values of the coefficients of the other two pairs, ν_2 and ν_3, which overlap each other, rise. Since the dilatancy is stretched in the direction of stretching, it is transverse for deformations of other directions. The coefficients of the first pair have the same values, $\nu_{12} = \nu_{13}$, but the coefficients of the other two pairs, $\nu_2$and ν_3, differ significantly. The graphs that make up the second pair of coefficients, ν_3 and ν_{23}, reveal their behavior—the values of ν_{23}, exceed the number 0.5.

Figure 3b shows graphs of the dependence of transverse deformations during compression. The line shown by the dots refers to the main direction coinciding with the voltage σ_3, and the graphs with the symbols ν_c and $\nu_3$should be considered the main "source" of dilatancy. As they increase, they cross the value of 0.5, which is typical for many loosening materials. The graphs for the coefficients with the symbols ν_1 and ν_2 coincide, slightly deviating from the graph for the curve $\Sigma\nu_i/3$, although the curves that make up them, ν_{21} and ν_{23}, are almost symmetrical.

The deformation anisotropy is more clearly shown on the graphs for the pliability of the bulk elasticity in the direction of the main stresses. The total volume deformation is determined by the formula (22), where $\phi_{ki} = \phi_k + æ(a + c_i)\Phi_d = 3(1 - \nu_i)/E_i-$ the compliance of the volume elasticity in the directions of the main stresses. In contrast to the theoretical volume compliance of ϕ_k the characteristics of ϕ_{ki} are smooth and continuous functions of stresses. Its first term is the pliability ϕ_k, established by the methodology, the second with a coefficient $a = 2k/3$, which is responsible for taking into account the dependence of the average voltage, and the third with a coefficient $c_{i.}$

which determines the directions of the axes. Give ϕ_{ki}, value (reverse module), to allow any state to find the values of three parameters changing of elasticity:

$$\vartheta_{\xi i} = \frac{\phi_k}{\phi_{\text{ki}}} = \frac{\text{K}_{\xi\text{i}}}{\text{K}_\xi}, \tag{33}$$

defining them as the degree of deviation from the theoretical volumetric compliance, which is the average, $\phi_k = \phi_{\text{ki}}/3$, for three compliance ϕ_{ki}. Each of them refers to the main stress, in the direction of which the initial values of the volume elasticity modules $\text{K}_{\xi\text{i}} = 1/\phi_{ki}$ are calculated (for $\sigma_i = 0$). In **Figure 4** curves 1, 2, and 3 represent graphs of these parameters $\vartheta_{\xi 1}$ по оси. The value of the parameter $\vartheta_{\xi 1}$ on curve 1 exceeds the values of other curves with a rapid decrease along the axis ξ to the value $\vartheta_0 = 1$. Judging by the shape of these curves, the elementary volume acquires the greatest deformation anisotropy in the direction with index 1. Curves with indices 2 and 3, having at first equal and small values compliance with the growth of the

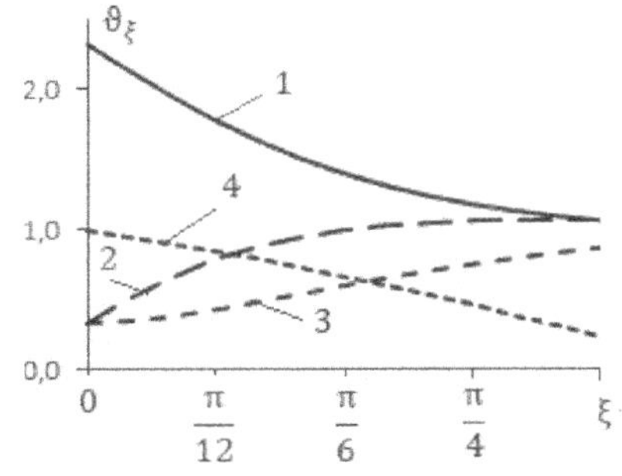

Figure 4.
Curves of changes in the values of the parameters of the changing elasticity $\vartheta_{\xi i}$.

angle ξ, increase slowly and in different ways. The third curve is affected by the presence of negative stress along the axis 3, given that the curves for these coefficients of transverse deformations overlap each other. Curve 4, denoted by the symbol $\vartheta_{\xi}^{*} = \phi_k/\phi_{kmax}$, is the ratio of the pliability of $\phi_k = \phi_{ki}/3$ to its first value.

The behavior of the curves for the parameters $\vartheta_{\xi i}$ can be associated with the behavior of the interstructural connections involved in creating the dilatancy for each state. The ordinates of the points of the curves, as it were, show the number of lost connections related to dilatancy. Numerical values of parameters can be useful for comparing the behavior of different materials, which is an important procedure for their analysis and practical selection of materials that differ, for example, in the binding matrix. At the same time, the material more clearly exerts a real deformation anisotropy [16, p. 151].

Briefly still on the shape change, it should be noted that the initial values of shear moduli $G_{\xi i}$ or their yields $\Phi_{\xi i}$ during the shape change deformation have no such features as the bulk yields, although the different values of the transverse strain ratios ν_{ij}, analyzed above, naturally influence their behavior. Nevertheless, the ratios of the strain intensities found, as from the initial data associated with the experimental results, to the strain intensities found after the matrix transformation of Eq. (19) are equal to 1. The high accuracy of each strain is especially valuable in determining the Lode parameters [15] when processing the results of experimental studies carried out in the 30–50 years of the last century. In order to estimate the nonlinear properties of the materials used, researchers resorted to constructing Lode diagrams based on the results of experimental studies, for example, in [17–19] by testing tubular specimens. In the test process, two strains are most often measured: axial and circumferential. And the researchers had to calculate the radial strain from the condition of "incompressibility," considering the sum of these three strains equal to zero. This led to a noticeable discrepancy in the results of each author, so that the author of the already quoted book [1] placed in it a diagram of the S-shaped curve with a minimum and a maximum.

The solution to this problem is formulated using tensor-nonlinear Equations [15]. Using the material functions of the proposed equations, finding the difference of Lode parameters, $\Delta\lambda = \lambda_\sigma - \lambda_\varepsilon$, without assumptions, diagrams with one minimum were obtained. The first λ_σ for stresses and the second for deformations:

$$\lambda_\varepsilon = 3(\varepsilon_2 - \varepsilon_1 - \varepsilon_3)/(\varepsilon_1 - \varepsilon_3), \quad (34)$$

where the former repeats the same fraction with the principal stresses by which it is determined. The problem of the researchers was to determine λ_ε.

9. Conclusion

A variant of the tensor-nonlinear equations, which can become the main direction in the nonlinear theory of elasticity, is proposed for wide use. This concept leads to taking into account dilatancy and strain anisotropy, about which Novozhilov V.V. prudently expressed in his work. They were used to study the properties of different-module materials and show that this mathematical apparatus is suitable not only for describing second-order smallness effects but also for describing effects associated with changes in the material structure. The influence of dilatancy on all the characteristics of form change and bulk elasticity is revealed, since its development with proportional stress growth is the main cause of deformation anisotropy, both of transverse strain coefficients and of bulk elasticity yields (or modules), which are directly related to the changing elasticity parameter, which is a quantitative estimate of these changes. In tensile and near-tensile states, its values significantly exceed unity. This can be explained by the fact that, in the first direction, dilatancy, being transverse for the other directions, causes transverse strain coefficients with values exceeding the number 0.5. The assumption of dilatancy to elastic deformations is an unavoidable step to trace the behavior of all deformations along the three directions. The exact coincidence of the total bulk strain as the sum of its components in the direction of the principal stresses, or, as the sum of linear-elastic and dilatancy, indicates recognition of the fact that the apparatus of the proposed equations may be a major trend in nonlinear elasticity theory. Whatever concepts other elasticity theories may adhere to, taking into account the real values of transverse strain coefficients in tension and compression will implicitly lead to the consideration of dilatancy and, consequently, to the difference in the values of the bulk elasticity characteristics. The next stage in the development of the nonlinear theory of elasticity is the involvement of the apparatus of thermodynamics.

Author details

Kirill F. Komkov
Technical Sciences, Military Technical University of Balashikha, Russia

*Address all correspondence to: 06kfk38@mail.ru

References

[1] Bel JF. Experimental Foundations of Mechanics of Deformable Solids. Moscow: Nauka; 1984. pp. part I-596c-part II-431c

[2] Rainer M. Mathematical theory of dilatancy. American Journal of Mathematics. 1945;**67**:350-362

[3] Novozhilov VV. On the relationship between stresses and deformations in a nonlinear elastic medium. Izvestiya Akademii Nauk SSSR. 1951;**15**(2): 183-194

[4] Novozhilov VV. On the principles of processing the results of static tests of isotropic materials. Applied Mathematics and Mechanics. 1951;**XV**: 709-722

[5] Lurie AI. Nonlinear Theory of Elasticity. Nauka: Moscow; 1980. p. 512

[6] Matchenko NM, Tolokonnikov LA. On the relationship of stresses with deformations in multi-modulus isotropic media. Izvestiya RAN, Mekhanika Tverdogo Tela. 1968;**6**: 108-110

[7] Komkov KF. On the methodology for determining the volume elasticity modulus and parameters that take into account loosening and changing the elasticity of composites based on tensor-nonlinear equations. Izvestiya RAN, Mekhanika Tverdogo Tela. 2019;**1**:50-62

[8] Komkov KF. Features of elastic properties of highly filled polymer materials. Bulletin of the Bauman Moscow State Technical University. Ser. Mashinostroenie. 2008;**3**(72):3-13

[9] Leonov MY, Ponyaev VA, Rusinko KN. Dependence Between Deformations and Stresses for Semi-Brittle Bodies. MTT. 1967;**6**:26-32

[10] Komkov KF. Description of anisotropy of isotropic materials caused by plastic deformation. Izvestiya RAN, Mekhanika Tverdogo Tela. 2008;**1**: 147-153

[11] Schwartz FR. On the mechanical properties of unfilled and filled elastomers. Mechanics and chemistry of sold fuels. In: Proceedings of the Fourth Symposium on Marine Structural Mechanics. Lafayette, Indiana: Purdue University; 1965. pp. 19-21

[12] Moshev VV. Structural Mechanics of Granular Composites on an Elastomeric Basis. Moscow: Nauka Publishing House; 1992. p. 79

[13] Mullins L. Softening of rubber by deformation, rub. Chemical Technology. 1969;**42**(1):339-362

[14] Korn G, Korn T. Handbook of Mathematics. Moscow: Nauka; 1977. p. 831

[15] Komkov KF. On the use of tensor-nonlinear equations for analyzing the behavior of plastic media. Bulletin of the Bauman Moscow State Technical University. Ser. Mechanical Engineering. 2007;**1**:46-56

[16] Novozhilov VV. Theory of elasticity. L.: Sudpromgiz, 1958. 370 p

[17] Lode U. Versuche uber den Einfußs der mittltren Hauptspannung auf das Fließen der Metalle Eisen, Kupfer and Nickel. Physicist. 1926;**36**:913-939

[18] Davis Evan A. An increase in stress at constant strain, and the relationship of stress–strain in the plastic state for copper in combined stress. Journal of Applied Mechanics. 1943;**10**:A-187-A-196

[19] Komkov KF. To the determination of Code parameters when processing test results Izv. Academy of Sciences of the Ukrainian SSR. MTT. 2005;**2**:126-135

Chapter 4

Obtaining of a Constitutive Models of Laminate Composite Materials

Mario Acosta Flores, Eusebio Jiménez López and Marta Lilia Eraña Díaz

Abstract

The study of the mechanical behavior of composite materials has acquired great importance due to the innumerable number of applications in new technological developments. As a result, many theories and analytical models have been developed with which its mechanical behavior is predicted; these models require knowledge of elastic properties. This work describes a basic theoretical framework, based on linear elasticity theory and classical lamination theory, to generate constitutive models of laminated materials made up of orthotropic layers. Thus, the models of three orthotropic laminated composite materials made up of layers of epoxy resin reinforced with fiberglass were also obtained. Finally, by means of experimental axial load tests, the constants of the orthotropic layers were determined.

Keywords: composite materials, elasticity theory, orthotropic materials, experimental methods, Sheet theory

1. Introduction

One of today's engineering needs is to develop new materials capable of improving the common materials that exist today (such as metals), in weight, wear resistance, corrosion resistance, high strength and stability at high temperatures, among others [1, 2]. The properties are improved through the use of reinforcements with fibers or particles in polymers, metals and ceramics, among others, giving rise to composite materials. The uses of composite materials can be found in the automotive industry, in the wind, aerospace and military industries, in civil applications, among others [3, 4]. The mechanical behavior of CM in tension, bending, torsion, etc., have been studied for decades [5–7]. For example, Sun [8] used glass fiber reinforced polyester resin to improve mechanical properties such as tensile strength, flexural strength, and Young's Modulus for single and multiple fibers. Acosta [9] developed a novel method to determine the stresses in torsion problems of laminated trimetallic and bimetallic composite bars, for which experimental and numerical analysis were carried out.

On the other hand, a necessary task for engineering applications is obtaining the mechanical properties of composite materials such as Young's Modulus (E), Rigidity Modulus (G), Yield Stress and Maximum Stress at traction, among others. In this regard, various authors have developed various numerical models and experimental techniques (photo-acoustic, ultrasound, Moiré interferometry, electrical extensometry, etc.) which have been applied to the design of composite materials [10, 11]. To obtain the effective properties of composite materials with different configurations, the authors Acosta et al. [12], developed an analytical constitutive model that is used for the mechanical analysis of intralaminar and global stresses in laminated composite materials with isotropic plies subject to axial load and to determine the elastic constants (E, v, and G)) of each of its components, using the method of electrical extensometry.

Of the laminated composite structures, the most widely used are those formed by layers of orthotropic materials. The design and mechanical analysis of laminated composite material structures involves a large number of variables (fiber orientation, layer thickness and stacking sequence, material densities, topological design, etc.) [13]. Of the laminated composite structures, the most widely used are those formed by orthotropic layers.

The study of the mechanical behavior of laminate composite materials is of great importance for engineering, so it is necessary to have a theoretical framework for its analysis, both globally and locally. In this work, the conceptual and analytical models foundations of the theory of linear elasticity and of the classical theory of laminated composite materials are presented for the theoretical and experimental approach of models that predict the mechanical behavior and allow obtaining its effective mechanical properties of a multi directionally reinforced laminated composite by orthotropic layers reinforced with longitudinal fibers [14–17]. With the models, the real properties obtained imply the effects of the existing defects in the interfaces between the layers (glue, gluing defects, layer fusion, etc.), which should considerably improve the efficiency in stress analysis.

2. Theory of elasticity

Stresses are internal forces that occur in bodies as a result of applying forces on their boundaries. If an imaginary cut is made in a body and if the internal distribution of forces on the cut surface is analyzed, then the stresses can be obtained as follows:

$$\sigma_{ij} = \underset{\Delta A \to 0}{Lim} \frac{\Delta F}{\Delta A} \tag{1}$$

where i and j are the Cartesian components x, y or z. For this case, i corresponds to the normal to the imaginary plane analyzed and j is associated with the direction of the force ΔF applied on the element of area ΔA. The state of stresses at a point can be represented graphically, as shown in **Figure 1**.

The stresses that act normally to the surface are called normal stresses $(\sigma_x, \sigma_y, \sigma_z)$, while those forces that act tangent to the surface are known as shearing stresses $(\tau_{xy}, \tau_{xz}, \tau_{yz})$. The complete stress analysis in a body implies determining the state of stresses in each of the points that make it up, and a partial analysis, in one or a set of points sufficient to solve a particular problem. The stress model is continuous and linear, from the mathematical point of view, which implies working with neighborhoods of infinitesimal points. By applying static equilibrium, the following field or equilibrium equations are obtained [14]:

$$\frac{\partial \sigma_x}{\partial x} + \frac{\partial \tau_{yx}}{\partial y} + \frac{\partial \tau_{zx}}{\partial z} + F_x = 0 \tag{2}$$

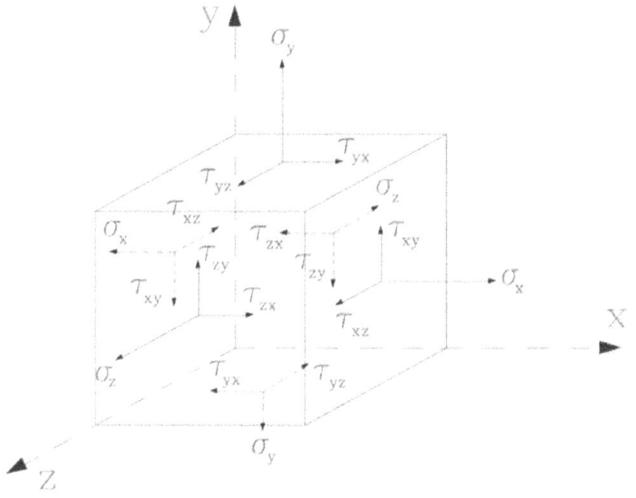

Figure 1.
State of stresses on a point.

$$\frac{\partial \tau_{xy}}{\partial x}+\frac{\partial \sigma_y}{\partial y}+\frac{\partial \tau_{zy}}{\partial z}+F_y=0$$

$$\frac{\partial \tau_{xz}}{\partial x}+\frac{\partial \tau_{yz}}{\partial y}+\frac{\partial \sigma_z}{\partial z}+F_z=0$$

On the other hand, a strain is defined as the relative displacement between the internal points of a body. If we consider a change in length in a straight-line segment in the x, y and z axes, we have normal or longitudinal strain ε_x, ε_y y ε_z. In the same way, if we have a transformation between the angles of two straight lines, the shear, or angular, strains γ_{xy}, γ_{xz}, y γ_{yz}, are obtained in the xy, xz and yz planes, respectively. The following expression represents the strain–displacement equations for normal and shear strains:

$$\varepsilon_{ij}=\frac{\partial u_i}{\partial x_j}+\frac{\partial u_j}{\partial x_i} \tag{3}$$

Or, explicitly:

$$\varepsilon_x=\frac{\partial u}{\partial x};\varepsilon_y=\frac{\partial v}{\partial y};\varepsilon_z=\frac{\partial w}{\partial z}; \tag{4}$$

$$\gamma_{xy}=\frac{\partial u}{\partial y}+;\gamma_{xz}=\frac{\partial u}{\partial z}+\frac{\partial w}{\partial x};\gamma_{yz}=\frac{\partial v}{\partial z}+\frac{\partial w}{\partial y}$$

The strain model described in expressions (Eq. (3)) is considered linear and continuous, which implies a model of infinitesimal strains. According to (Dally), the linear equations between stresses and strains give rise to the constitutive model. These equations are determined according to the following expression:

$$\sigma_i=C_{ij}\varepsilon_j; i,j=1,2,3,\ldots,6 \tag{5}$$

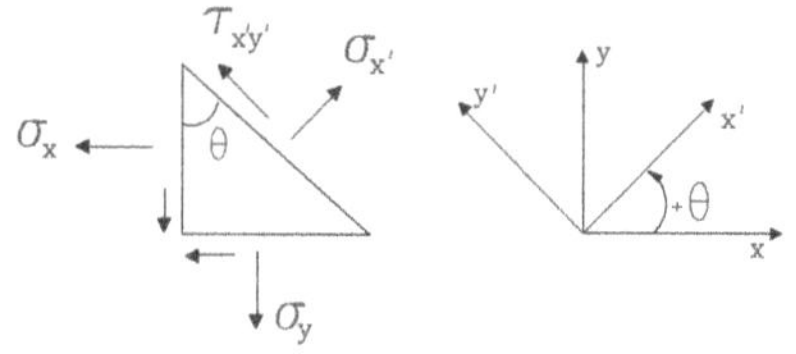

Figure 2.
Transformation of stresses at a point with state of plane stresses.

where σ_i are the stress components, C_{ij} is the stiffness tensor and ε_j are the strain components. The explicit form of expression (Eq. (5)) is known as Hooke's Generalized Law. This is:

$$\begin{aligned}
\sigma_x &= C_{11}\varepsilon_x + C_{12}\varepsilon_y + C_{13}\varepsilon_z + C_{14}\gamma_{xy} + C_{15}\gamma_{yz} + C_{16}\gamma_{zx} \\
\sigma_y &= C_{21}\varepsilon_x + C_{22}\varepsilon_y + C_{23}\varepsilon_z + C_{24}\gamma_{xy} + C_{25}\gamma_{yz} + C_{26}\gamma_{zx} \\
\sigma_z &= C_{31}\varepsilon_x + C_{32}\varepsilon_y + C_{33}\varepsilon_z + C_{34}\gamma_{xy} + C_{35}\gamma_{yz} + C_{36}\gamma_{zx} \\
\tau_{xy} &= C_{41}\varepsilon_x + C_{42}\varepsilon_y + C_{43}\varepsilon_z + C_{44}\gamma_{xy} + C_{45}\gamma_{yz} + C_{46}\gamma_{zx} \\
\tau_{yz} &= C_{51}\varepsilon_x + C_{52}\varepsilon_y + C_{53}\varepsilon_z + C_{54}\gamma_{xy} + C_{55}\gamma_{yz} + C_{56}\gamma_{zx} \\
\tau_{xz} &= C_{61}\varepsilon_x + C_{62}\varepsilon_y + C_{63}\varepsilon_z + C_{64}\gamma_{xy} + C_{65}\gamma_{yz} + C_{66}\gamma_{zx}
\end{aligned} \tag{6}$$

Here, $C_{11}, C_{12}, \ldots C_{66}$ (C_{ij} for $i,j = 1, 2, \ldots, 6$), are called the stiffness constants of the material and are independent of the stress values or the strain values. The following expression shows the inverse form between strains and stresses:

$$\varepsilon_i = S_{ij}\sigma_j, i,j = 1, 2, 3, \ldots, 6 \tag{7}$$

Here, S_{ij} is known as the compliance tensor. According to Durelli [14] and when considering the strain energy in the analysis, the following expressions are fulfilled:

$$C_{ij} = C_{ji}; S_{ij} = S_{ji} \tag{8}$$

It is worth mentioning that the constants of the constitutive models can be put as a function of the so-called engineering constants (Young's modulus (E), Poisson's ratios (ν) and shear modulus (G)), which can be obtained through tests of pure tension and shear. According to Durelli [14] the model of the Theory of Linear Elasticity that governs the bodies′ mechanical analysis, is composed of a system of 15 partial differential equations and 15 unknowns (σ_{ij}, $\varepsilon_{ij} y u_i$). On the other hand, for practical purposes, it is necessary to characterize the state of stress at a body's point on an arbitrary plane. The analytical equations that govern the state of stress at a body's point that make it possible to linearly transform the stress components, refer to a reference coordinate system and find the stress components regarding any other system. **Figure 2** shows a graphic example of a state of plane stresses.

In the case of the strain transformation laws, a similar process is carried out.

3. Mechanical analysis in orthotropic materials

An orthotropic material is one in which the values of its elastic properties are different for each orientation, referred to three coordinate axes, each perpendicular to another (see **Figure 3**). Examples of orthotropic materials are: wood, unidirectionally materials reinforced with fiberglass, carbón, Kevlar, among others. In orthotropic materials, the stiffness changes depending on the orientation of the

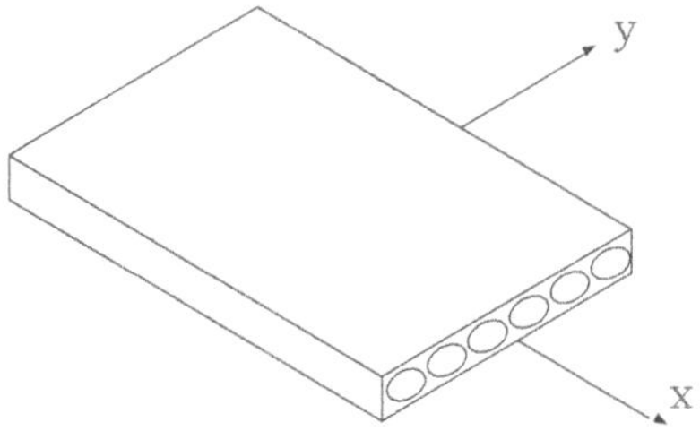

Figure 3.
Axes of symmetry of a plane orthotropic material.

fibers. To determine the stress or strain components in any direction, it is necessary to know the states of stress or strain at a point and apply the analytical transformation equations. The stress–strain relations and the stress and strain transformation equations are the basis for the construction of constitutive models to study the stresses and to determine the effective mechanical properties of laminated composite materials as a function of the orientation and direction [16].

To clarify the analysis, the *xy* system will be used when it coincides with the axes of symmetry of the properties of the unidirectional material and the system of 12 will be used for any coordinate system outside of it (see **Figure 4**).

The equations that govern the transformation of plane stresses, in a unidirectionally reinforced laminate, allow the obtention of the value of the stresses in the xy system once the stresses in the 12 system are known (see **Figures 5** and **6**).

To know the relations between the stresses of the *xy* system regarding the stresses in the 12 system, a cut perpendicular to the fibers is analyzed and the director cosines, m = cos θ and n = sin θ, are used (see **Figure 6**). These equations are expressed as follows:

$$\begin{aligned} \sigma_x &= \sigma_1 m^2 + \sigma_2 n^2 + \sigma_6 mn_6; \sigma_y = \sigma_1 n^2 + \sigma_2 m^2 - 2\sigma_6 mn; \\ \sigma_s &= -\sigma_1 mn + \sigma_2 mn + \sigma_6 [m^2 - n^2] \end{aligned} \tag{9}$$

Here, σ_x and σ_y are the normal stresses and σ_s corresponds to the shear stress in the xy system; σ_1 and σ_2 are the normal stresses and σ_6 is the shear stress in the 12 system.

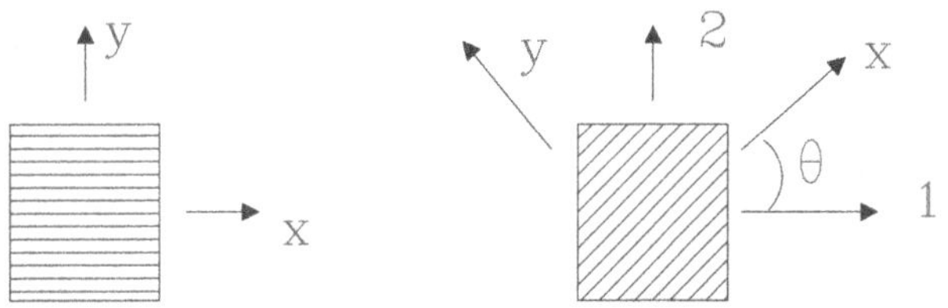

Figure 4.
System xy *in the direction of the fibers and system 12 in a different orientation.*

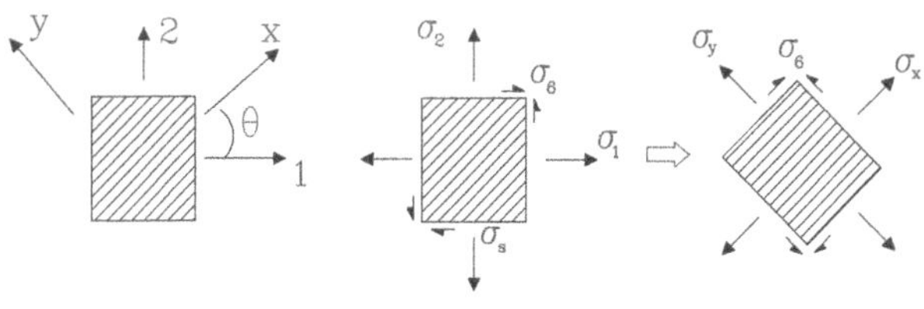

Figure 5.
Transformation of stress.

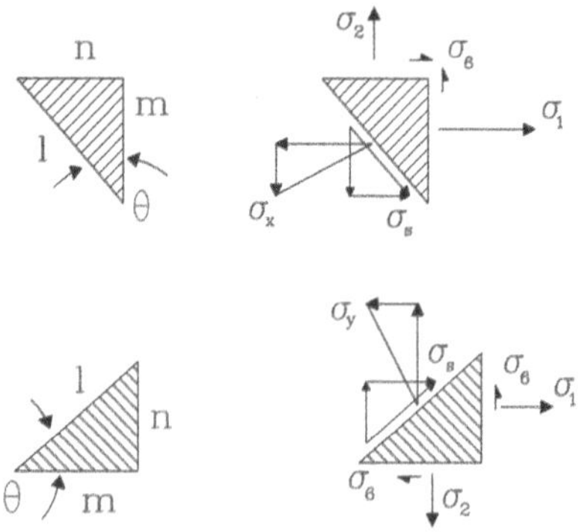

Figure 6.
Stress components in system xy *as a function of stress components in system* 12.

The strain transformation equations, between the xy Cartesian strains and the strains in system 12, in terms of the director cosines are shown in equations (Eq. (10)).

$$\begin{aligned} \varepsilon_x &= \varepsilon_1 m^2 + \varepsilon_2 n^2 + \varepsilon_6 m n_6; \varepsilon_y = \varepsilon_1 n^2 + \varepsilon_2 m^2 - \varepsilon_6 mn \\ \varepsilon_s &= -\varepsilon_1 mn + \varepsilon_2 mn + \varepsilon_6 \left[m^2 - n^2\right] \end{aligned} \tag{10}$$

Here, ε_x and ε_y are the normal strains and ε_s corresponds to the shear strain in the xy system; ε_1 and ε_2 are the normal strains and ε_6 is the shear strain in the 12 system. The linear relations, between stresses and strains in a plane system for an orthotropic composite, are obtained by means of Hooke's generalized law. These relations are as follows:

$$\begin{aligned} \sigma_x &= C_{11}\varepsilon_x + C_{12}\varepsilon_y + C_{13}\varepsilon_z; \sigma_y = C_{21}\varepsilon_x + C_{22}\varepsilon_y + C_{23}\varepsilon_z; \\ \sigma_z &= C_{31}\varepsilon_x + C_{32}\varepsilon_y + C_{33}\varepsilon_z \ \tau_{xy} = C_{44}\gamma_{xy}; \tau_{yz} = C_{55}\gamma_{yz}; \tau_{xz} = C_{66}\gamma_{zx} \end{aligned} \tag{11}$$

It is worth mentioning that the state of stress at all points is considered plane stress because it is assumed that the distribution of strains is homogeneous through the thickness of the orthotropic composite. The planes of symmetry correspond to the longitudinal direction of the fibers and the transverse direction, respectively. The composite material and its symmetry planes are shown in **Figure 3**. The material stiffness coefficients are obtained from the development of the following simple axial tests (see **Figure** 7): 1) Tension test in the longitudinal direction of the fibers, 2) Tension test in the cross-fiber direction and 3) Pure shear test.

For a uniaxial state of stress, we have the equations $\varepsilon_x = S_{11}\sigma_x$ and $\varepsilon_y = S_{21}\sigma_y$. By Hooke's law, the stress in the direction of the applied load P is: $\sigma = \frac{P}{A}$, if and only if it is assumed that the P force is applied uniformly to a cross section and the change of the latter for any P value is negligible [14]. The equations between the stress σ and the strain Ɛ (in the direction of σ) within the elastic-linear range is: $\sigma = \varepsilon E$, where E is the Young's modulus of the material. Note that if $S_{11} = \frac{1}{E_x}$, then

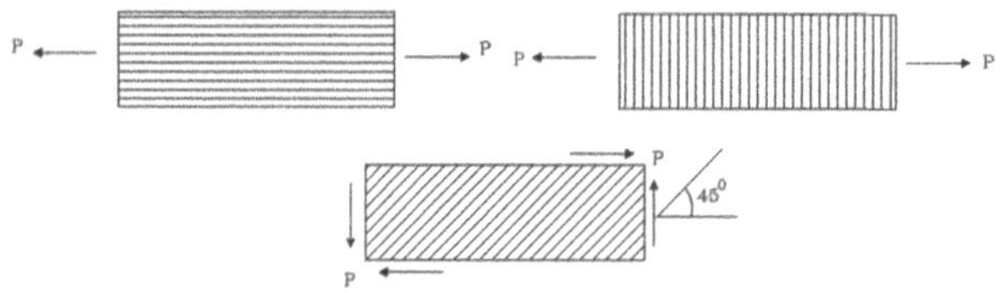

Figure 7.
Tests, tension in the direction of the fibers, tension in the direction transverse to the fibers and pure shear at 45°.

$$\varepsilon_x = \frac{\sigma_x}{E_x};\ \varepsilon_y = -\frac{\upsilon_x \sigma_x}{E_x} \tag{12}$$

Here, $\nu_x = -\frac{\varepsilon_y}{\varepsilon_x}$ is the longitudinal Poisson's ratio and ε_x and ε_y are the longitudinal and transverse deformations, respectively.

By following a similar process performed in the previous test (see **Figure 1**.4), the following equations are obtained:

$$\varepsilon_x = -\frac{\nu_y \sigma_y}{E_y};\ \varepsilon_y = -\frac{\sigma_y}{E_y} \tag{13}$$

where $\nu_y = -\frac{\varepsilon_x}{\varepsilon_y}$ is the transverse Poisson's ratio, E_y is the transverse Young's Modulus and ε_y and ε_x are the longitudinal and transverse strains, respectively. In a pure shear test (see **Figure** 7), the constitutive relation between shear stress and shear strain is as follows:

$$\gamma_{xy} = \frac{\tau_{xy}}{G_{xy}} \tag{14}$$

Here, τ_{xy} is the shear stress, γ_{xy} is the shear strain, and G_{xy} is the shear modulus of the material. As result of the three tests, the following constitutive relations are obtained:

$$\varepsilon_x = \frac{\sigma_x}{E_x} - \frac{\nu_y \sigma_y}{E_y};\ \varepsilon_y = -\frac{\nu_x \sigma_x}{E_x} - \frac{\sigma_y}{E_y};\ \gamma_{xy} = \frac{\tau_{xy}}{G_{xy}} \tag{15}$$

The relations that define the stiffness constants as a function of the engineering constants are as follows:

$$Q_{xx} = \frac{E_x}{1-\upsilon_x\upsilon_y};\ Q_{yy} = \frac{E_y}{1-\upsilon_y\upsilon_x};\ Q_{yx} = \frac{\upsilon_x E_y}{1-\upsilon_x\upsilon_y};\ Q_{xy} = \frac{\upsilon_y E_x}{1-\upsilon_x\upsilon_y};\ Q_{ss} = E_{ss} \tag{16}$$

The relations between compliance constants and engineering constants are:

$$S_{xx} = \frac{1}{E_x};\ S_{yy} = \frac{1}{E_y};\ S_{yx} = \frac{\upsilon_y}{E_y};\ S_{xy} = \frac{\upsilon_x}{E_x} Q_{ss} = \frac{1}{G_{xy}} \tag{17}$$

By symmetry of the stiffness tensor and the compliance tensor, we have: $\frac{\upsilon_x}{E_x} = \frac{\upsilon_y}{E_y}$. This relation reduces the number of independent constants, from five to four, in the plane problem. To determine the stresses in terms of the strains (for a reference system), different from the material's symmetry axes, system 12 (see **figure 8**). Is necessary to apply a strain transformation process to change to the symmetry axes (system xy), configuration (a) to (b) and determine the state of stresses with constitutive model, configuration (b) to (c) to subsequently apply a stress transformation and come at the stresses in axes 12 (configuration (c) to (d)). To transform configuration (a) to (d), knowing the stiffness constants in the specific orientation and the value of the strain components in same direction, are the following steps:

Step 1) The step from configuration (a) to (b) is obtained by making a positive strain transformation and using equations (Eq. (10)), that is:

$$\begin{aligned} &\varepsilon_x = m^2\varepsilon_1 + n^2\varepsilon_2 + mn\varepsilon_6;\ \varepsilon_y = n^2\varepsilon_1 + m^2\varepsilon_2 - mn\varepsilon_6; \\ &\varepsilon_s = -2mn\varepsilon_1 + 2mn\varepsilon_2 + \left[m^2 - n^2\right]\varepsilon_6 \end{aligned} \tag{18}$$

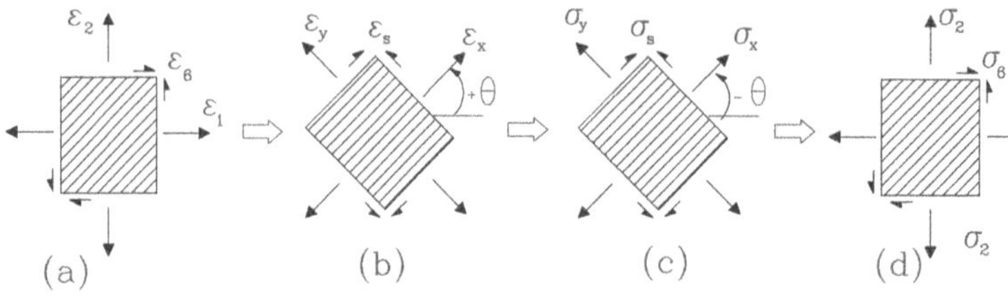

Figure 8.
Transformation process, from the state of strains to the state of stresses, in a coordinate system 12. Strain transformation process, (a) to (b), getting stresses with constitutive model, (b) to (c) and finally arrive at the state of stresses in axes 12, (c) to (d).

Step 2) To go from configuration (b) to configuration (c), the stress–strain relations are used in the material's symmetry axes. These relations are as follows:

$$\begin{aligned}\sigma_x &= Q_{xx}[m^2\varepsilon_1 + n^2\varepsilon_2 + mn\varepsilon_6] + Q_{xy}[n^2\varepsilon_1 + m^2\varepsilon_2 - mn\varepsilon_6];\\ \sigma_y &= Q_{yx}[m^2\varepsilon_1 + n^2\varepsilon_2 + mn\varepsilon_6] + Q_{yy}[n^2\varepsilon_1 + m^2\varepsilon_2 - mn\varepsilon_6];\\ \sigma_{ss} &= Q_{ss}[-2m\varepsilon_1 + 2mn\varepsilon_2 + [m^2 - n^2]\varepsilon_6]\end{aligned} \tag{19}$$

Step 3) To go from configuration c) to d), the stress transformation equations with negative angle of rotation are used. This is:

$$\begin{aligned}\sigma_1 = m^2 &\left[Q_{xx}\left[m^2\varepsilon_1 + n^2\varepsilon_2 + mn\varepsilon_6\right] + Q_{xy}\left[n^2\varepsilon_1 + m^2\varepsilon_2 - mn\varepsilon_6\right]\right]\\ &+ n^2\left[Q_{yx}\left[m^2\varepsilon_1 + n^2\varepsilon_2 + mn\varepsilon_6\right] + Q_{yy}\left[n^2\varepsilon_1 + m^2\varepsilon_2 - mn\varepsilon_6\right]\right]\\ &- 2mn\left[Q_{ss}\left[-2mn\varepsilon_1 + 2mn\varepsilon_2 + \left[m^2 - n^2\right]\varepsilon_6\right]\right]\\ \sigma_2 = n^2 &\left[Q_{xx}\left[m^2\varepsilon_1 + n^2\varepsilon_2 + mn\varepsilon_6\right] + Q_{xy}\left[n^2\varepsilon_1 + m^2\varepsilon_2 - mn\varepsilon_6\right]\right]\\ &+ m^2\left[Q_{yx}\left[m^2\varepsilon_1 + n^2\varepsilon_2 + mn\varepsilon_6\right] + Q_{yy}\left[n^2\varepsilon_1 + m^2\varepsilon_2 - mn\varepsilon_6\right]\right]\\ &+ 2mn\left[Q_{ss}\left[-2mn\varepsilon_1 + 2mn\varepsilon_2 + \left[m^2 - n^2\right]\varepsilon_6\right]\right]\\ \sigma_6 = mn &\left[Q_{xx}\left[m^2\varepsilon_1 + n^2\varepsilon_2 + mn\varepsilon_6\right] + Q_{xy}\left[n^2\varepsilon_1 + m^2\varepsilon_2 - mn\varepsilon_6\right]\right]\\ &- mn\left[Q_{yx}\left[m^2\varepsilon_1 + n^2\varepsilon_2 + mn\varepsilon_6\right] + Q_{yy}\left[n^2\varepsilon_1 + m^2\varepsilon_2 - mn\varepsilon_6\right]\right]\\ &+ (m^2 - n^2)\left[Q_{ss}\left[-2mn\varepsilon_1 + 2mn\varepsilon_2 + \left[m^2 - n^2\right]\varepsilon_6\right]\right]\end{aligned} \tag{20}$$

Considering the strains and the symmetry of the stiffness tensor, the constitutive relations can be obtained in an arbitrary orientation, system 12 (see **Figure 9**), to a direct transformation from the state of strain to the state of stress. What would be:

$$\begin{aligned}&\sigma_1 = Q_{11}\varepsilon_1 + Q_{12}\varepsilon_2 + Q_{16}\varepsilon_6; \sigma_2 = Q_{21}\varepsilon_1 + Q_{22}\varepsilon_2 + Q_{26}\varepsilon_6;\\ &\sigma_6 = Q_{61}\varepsilon_1 + Q_{62}\varepsilon_2 + Q_{66}\varepsilon_6\end{aligned} \tag{21}$$

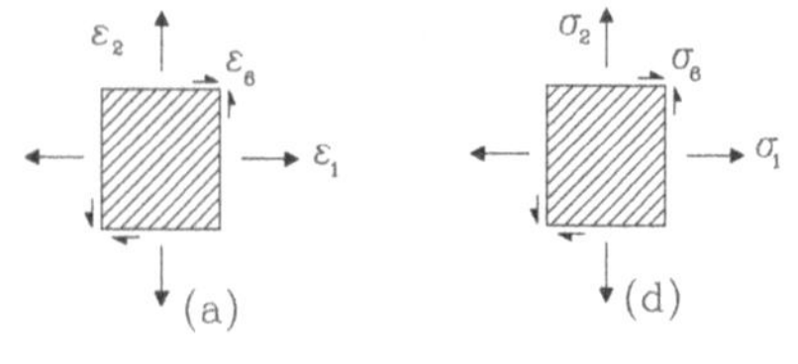

Figure 9.
Direct transformation from the state of strain to the state of stress in a system 12, getting the state of stresses in terms of the state of strains, with constitutive relations. Configuration a) to d).

The stiffness constants in an arbitrary orientation are defined as follows:

$$
\begin{aligned}
Q_{11} &= m^4 Q_{xx} + 2m^2n^2 Q_{xy} + n^4 Q_{yy} - 4m^2n^2 Q_{ss} \\
Q_{12} &= m^2n^2 Q_{xx} + (m^4 + n^4) Q_{xy} + m^2n^2 Q_{yy} - 4m^2n^2 Q_{ss} \\
Q_{16} &= m^3n Q_{xx} + (mn^3 + m^3n) Q_{xy} - mn^3 Q_{yy} + 4[mn^3 - m^3n] Q_{ss} \\
Q_{22} &= n^4 Q_{xx} + 2m^2n^2 Q_{xy} + m^4 Q_{yy} + 4m^2n^2 Q_{ss} \\
Q_{26} &= mn^3 Q_{xx} + (m^3n - mn^3) Q_{xy} - m^3n Q_{yy} + 2[m^3n - mn^3] Q_{ss} \\
Q_{66} &= m^2\, n^2 Q_{xx} - 2m^2n^2 Q_{xy} + m^2n^2 Q_{yy} + [m^2 - n^2]^2 Q_{ss}
\end{aligned}
\tag{22}
$$

The obtained results are of great importance due to these are the equations that define the variation of the constants as a function of the orientation. On the other hand, the constitutive relations and the compliance constants, in a different orientation from the axis of symmetry of the material, are obtained by applying a similar process to the one in the previous section. But now we start from the known state of stresses and is required to know the state of the strains. To process it, need to apply a stresses transformation process to change to the symmetry axes (system xy), configuration (a) to (b), determine the state of strains with constitutive model, configuration (b) to (c) to subsequently apply a strain transformation and get the strains in axes 12, configuration (c) to (d), see **Figure 10**. The relations between the constants of compliance, on the lines of symmetry and outside of them, are:

$$
\varepsilon_1 = S_{11}\sigma_1 + S_{12}\sigma_2 + S_{16}\sigma_6;\ \varepsilon_2 = S_{21}\sigma_1 + S_{22}\sigma_2 + S_{26}\sigma_6;\ \varepsilon_6 = S_{61}\sigma_1 + S_{62}\sigma_2 + S_{66}\sigma_6
\tag{23}
$$

The relations between the compliance constants, corresponding to the lines symmetry and outside of them, are:

$$
\begin{aligned}
S_{11} &= m^4 S_{xx} + 2m^2n^2 S_{xy} + n^4 S_{yy} + m^2n^2 S_{ss} \\
S_{12} &= m^2n^2 S_{xx} + (m^4 + n^4) S_{xy} + m^2n^4 S_{yy} - m^2n^2 S_{ss} \\
S_{16} &= 2m^3n S_{xx} + 2(mn^3 - m^3n) S_{xy} - 2mn^3 S_{yy} - mn(m^2 - n^2) S_{ss} \\
S_{22} &= n^4 S_{xx} + 2m^2n^2 S_{xy} + m^4 S_{yy} + m^2n^2 S_{ss} \\
S_{26} &= 2m^3n S_{xx} + 2(mn^3 - m^3n) S_{xy} - 2mn^3 S_{yy} - mn(m^2 - n^2) S_{ss} \\
S_{66} &= 4m^2n^2 S_{xx} - 8m^2n^2 S_{xy} + 4m^2n^2 S_{yy} + (m^2 - n^2)^2 S_{ss}
\end{aligned}
\tag{24}
$$

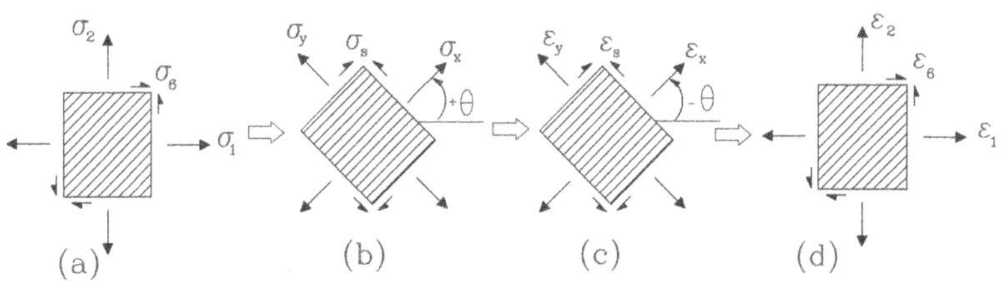

Figure 10.
Transformation process from the state of stresses to the state of strains in a coordinate system 12. Stresses transformation process, (a) to (b), getting strains with constitutive model, (b) to (c), and to finally obtain the state of strains in axes 12, (c) to (d).

Equations (Eq. (22)) can be put in terms of trigonometric identities. This is:

$$\begin{aligned}
Q_{11} &= \frac{1}{8}\left(3Q_{xx} + 2Q_{xy} + 3Q_{yy} + 4Q_{ss}\right) + \frac{1}{8}\left(4Q_{xx} - 4Q_{yy}\right)\cos 2\theta \\
&\quad + \frac{1}{8}\left(Q_{xx} - 2Q_{xy} + Q_{yy} - 4Q_{ss}\right)\cos 4\theta \\
Q_{12} &= \frac{1}{8}\left(Q_{xx} + 6Q_{xy} + Q_{yy} - 4Q_{ss}\right) - \frac{1}{8}\left(Q_{xx} - 2Q_{xy} + Q_{yy} - 4Q_{ss}\right)\cos 4\theta \\
Q_{16} &= \frac{1}{8}\left(2Q_{xx} - 2Q_{yy}\right)sen2\theta + \frac{1}{8}\left(Q_{xx} + Q_{yy} - 2Q_{xy} - Q_{ss}\right)sen4\theta \\
Q_{22} &= \frac{1}{8}\left(3Q_{xx} + 2Q_{xy} + 3Q_{yy} + 4Q_{ss}\right) - \frac{1}{8}\left(4Q_{xx} - 4Q_{yy}\right)\cos 2\theta \\
&\quad + \frac{1}{8}\left(Q_{xx} - 2Q_{xy} + Q_{yy} - 4Q_{ss}\right)\cos 4\theta \\
Q_{26} &= \frac{1}{2}\left(Q_{xx} - Q_{yy}\right)sen2\theta - \frac{1}{8}\left(Q_{xx} - 2Q_{xy} + Q_{yy} - 4Q_{ss}\right)sen4\theta \\
Q_{66} &= \frac{1}{8}\left(Q_{xx} - 2Q_{xy} + Q_{yy} - 4Q_{ss}\right) - \frac{1}{8}\left(Q_{xx} + Q_{yy} - 2Q_{xy} - Q_{ss}\right)sen4\theta
\end{aligned} \tag{25}$$

If and only if the following relations are satisfied:

$$\begin{aligned}
&m^4 = \frac{1}{8}(3 + 4\cos 2\theta + \cos 4\theta); m^3 n = \frac{1}{8}(2sen2\theta + sen4\theta); m^2 n^2 = \frac{1}{8}(1 - \cos 4\theta); \\
&mn^3 = \frac{1}{8}(2sen2\theta - sen4\theta); n^4 = \frac{1}{8}(3 - 4\cos 2\theta + \cos 4\theta)
\end{aligned} \tag{26}$$

By defining the following relations:

$$\begin{aligned}
&U_1 = \frac{1}{8}\left(3Q_{xx} + 2Q_{xy} + 3Q_{yy} + 4Q_{ss}\right); U_2 = \frac{1}{2}\left(Q_{xx} - Q_{yy}\right); \\
&U_3 = \frac{1}{8}\left(Q_{xx} - 2Q_{xy} + Q_{yy} - 4Q_{ss}\right); U_4 = \frac{1}{8}\left(Q_{xx} + 6Q_{xy} + Q_{yy} - 4Q_{ss}\right); \\
&U_5 = \frac{1}{8}\left(Q_{xx} - 2Q_{xy} + Q_{yy} - 4Q_{ss}\right)
\end{aligned} \tag{27}$$

And, when ordering terms, the following relatios are obtained:

$$\begin{aligned}
&Q_{11} = U_1 + U_2\cos 2\theta + U_3\cos 4\theta; Q_{12} = U_4 - U_3\cos 4\theta; Q_{16} = \frac{1}{2}U_2 sen2\theta + U_3 sen4\theta \\
&Q_{22} = U_1 - U_2\cos 2\theta + U_3\cos 4\theta; Q_{26} = \frac{1}{2}U_2 sen2\theta - U_3 sen4\theta; Q_{66} = U_5 - U_3\cos 4\theta
\end{aligned} \tag{28}$$

4. Laminate composite materials theory

A laminate is a set of plies or ply groups that have different orientations from their main axes [16]. The classical laminate theory assumes, in the mechanical model, the following [16, 17]: the laminate is symmetric; the behavior of the plies and the laminate complies with Hooke's law; each ply is considered orthotropic; the union between plies is perfect and thin; the functions of the displacements and strains are considered continuous through the interface; the laminate is homogeneous, elastic and linear; and the ply thicknesses are constant, thin and homogeneous throughout the laminate.

For the study of global stresses, the following aspects are assumed: the model is linear [14]; and the state of stress is homogeneous throughout the laminate. Thus, edge effects on the laminate can be ignored, allowing the problem to be about plane stresses. For the stress analysis at the local level, it is assumed that the problem for each of the plies is biaxial of stresses, and that the normal stresses have an average constant distribution through the thickness of the plies (see **Figure 11**). On a global level at a symmetric laminate: it is made up of homogeneous plies; the union between plies is perfect; and the laminate's composite thickness is homogeneous. Finally, when considering a state of homogeneous strain in the laminate and in the plies, the intralaminar stresses τ_{yz} can be ignored, so that all the points in the laminate and locally in the plies present a state of plane stresses.

A laminate composite material can be defined by means of a code [16]. **Figure 12** shows a diagram of a symmetrical laminate. Its code is $\left[45^0_3/90^0_1/0^0_2\right]_S$ and it is interpreted as follows: if the analysis is started from z = −h / 2, we have three plies oriented at 45^0, followed by a ply with orientation 90^0, and finally two plies at 0^0. The subscript S means that the laminate is symmetric and that from the central axis up, the sequence is in reverse order. If instead of S there were the subscript T, this would mean that the code would be written in full, that is: $\left[45^0_3/90^0_1/0^0_2/0^0_2/90^0_1/45^0_3\right]_T$. If the laminate were not symmetric, we would have a code like the following: $\left[45^0_3/90^0_1/0^0_2/45^0_3/90^0_1/0^0_2\right]_T$.

4.1 Mechanics of symmetric laminate

In the study of symmetric laminates, the strains in the *xy* plane are constant throughout the lamina if and only if its thickness is small compared to the length

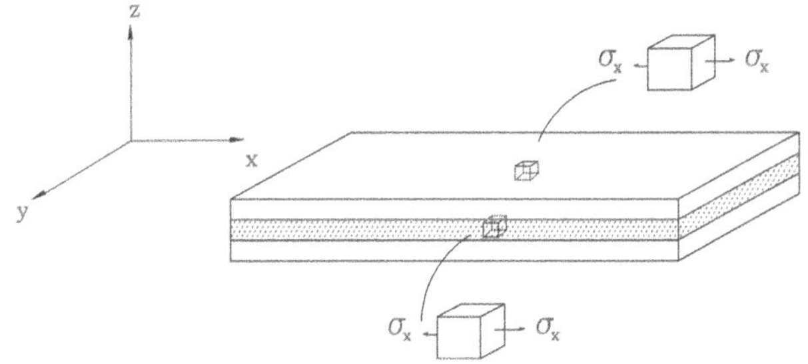

Figure 11.
Intralaminar stress state for a three-ply laminate.

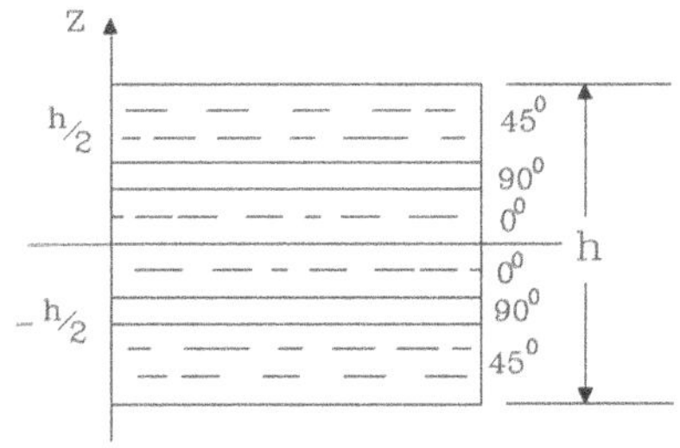

Figure 12.
Sequence of a symmetric laminate.

and width. Therefore, $\varepsilon_1(z) = \varepsilon_1^0$, $\varepsilon_2(z) = \varepsilon_2^0$ and $\varepsilon_{12}(z) = \varepsilon_{12}^0$. The exponent (0) means that the strains as a function of z are constant. It is worth mentioning that the stress distribution is not constant and varies from plies or ply group to plies. If a global analysis of the laminate is carried out, the constitutive relations are obtained based on the properties and orientation in each ply group [16]. For this study it is necessary to start from the concept of average effort, (see **Figure 13**). This is:

$$\overline{\sigma}_1 = \frac{1}{h}\int_{-\frac{h}{2}}^{\frac{h}{2}} \sigma_1 dz; \overline{\sigma}_2 = \frac{1}{h}\int_{-\frac{h}{2}}^{\frac{h}{2}} \sigma_2 dz; \overline{\sigma}_6 = \frac{1}{h}\int_{-\frac{h}{2}}^{\frac{h}{2}} \sigma_6 dz \tag{29}$$

On the other hand, the stresses are defined as a function of the stiffness constants in any direction. The stress–strain relations are as follow:

$$\begin{aligned} &\sigma_1 = Q_{11}\varepsilon_1 + Q_{12}\varepsilon_2 + Q_{16}\varepsilon_6; \sigma_2 = Q_{21}\varepsilon_1 + Q_{22}\varepsilon_2 + Q_{26}\varepsilon_6; \\ &\sigma_6 = Q_{61}\varepsilon_1 + Q_{62}\varepsilon_2 + Q_{66}\varepsilon_6 \end{aligned} \tag{30}$$

If the strains are constant, then the average stresses are expressed as follows:

$$\begin{aligned} \overline{\sigma}_1 = \frac{1}{h}\int_{-\frac{h}{2}}^{\frac{h}{2}} (Q_{11}\varepsilon_1^0 + Q_{12}\varepsilon_2^0 + Q_{13}\varepsilon_6^0) dz; \overline{\sigma}_2 = \frac{1}{h}\int_{-\frac{h}{2}}^{\frac{h}{2}} (Q_{21}\varepsilon_1^0 + Q_{22}\varepsilon_2^0 + Q_{23}\varepsilon_6^0) dz \\ \overline{\sigma}_6 = \frac{1}{h}\int_{-\frac{h}{2}}^{\frac{h}{2}} (Q_{61}\varepsilon_1^0 + Q_{62}\varepsilon_2^0 + Q_{66}\varepsilon_6^0) dz \end{aligned} \tag{31}$$

Considering that the constants Q vary from ply to ply, the average stresses take thefollowing form:

$$\begin{aligned} \overline{\sigma}_1 &= \frac{1}{h}\left[\int_{-\frac{h}{2}}^{\frac{h}{2}} Q_{11} dz \varepsilon_1^0 + \int_{-\frac{h}{2}}^{\frac{h}{2}} Q_{12} dz \varepsilon_2^0 + \int_{-\frac{h}{2}}^{\frac{h}{2}} Q_{13} dz \varepsilon_6^0\right]; \\ \overline{\sigma}_2 &= \frac{1}{h}\left[\int_{-\frac{h}{2}}^{\frac{h}{2}} Q_{21} dz \varepsilon_1^0 + \int_{-\frac{h}{2}}^{\frac{h}{2}} Q_{22} dz \varepsilon_2^0 + \int_{-\frac{h}{2}}^{\frac{h}{2}} Q_{23} dz \varepsilon_6^0\right] \\ \overline{\sigma}_6 &= \frac{1}{h}\left[\int_{-\frac{h}{2}}^{\frac{h}{2}} Q_{61} dz \varepsilon_1^0 + \int_{-\frac{h}{2}}^{\frac{h}{2}} Q_{62} dz \varepsilon_2^0 + \int_{-\frac{h}{2}}^{\frac{h}{2}} Q_{66} dz \varepsilon_6^0\right] \end{aligned} \tag{32}$$

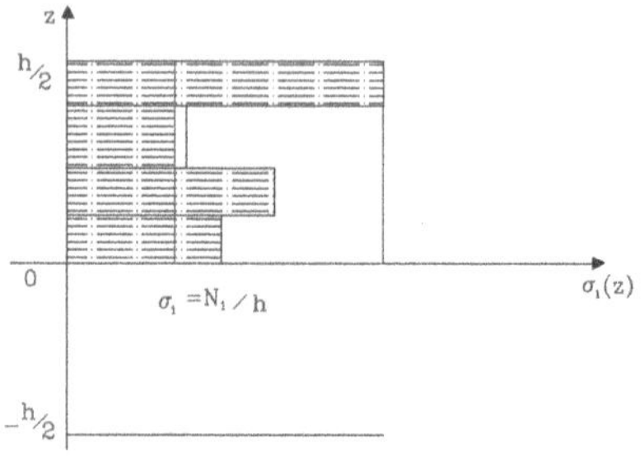

Figure 13.
Representation of mean stresses in a multidirectional lamina.

If:

$$A_{11} = \int_{-\frac{h}{2}}^{\frac{h}{2}} Q_{11}dz; A_{21} = \int_{-\frac{h}{2}}^{\frac{h}{2}} Q_{21}dz; A_{22} = \int_{-\frac{h}{2}}^{\frac{h}{2}} Q_{22}dz; A_{61} = \int_{-\frac{h}{2}}^{\frac{h}{2}} Q_{61}dz;$$
$$A_{62} = \int_{-\frac{h}{2}}^{\frac{h}{2}} Q_{62}dz; A_{66} = \int_{-\frac{h}{2}}^{\frac{h}{2}} Q_{66}dz \quad (33)$$

And: $A_{21} = A_{12}$; $A_{61} = A_{16}$ y $A_{62} = A_{26}$ (The equivalent modulus tensor A_{ij} is symmetric), the average stresses are rewritten as follows:

$$\overline{\sigma}_1 = \frac{1}{h}\left[A_{11}\varepsilon_1^0 + A_{21}\varepsilon_2^0 + A_{61}\varepsilon_6^0\right]; \overline{\sigma}_2 = \frac{1}{h}\left[A_{21}\varepsilon_1^0 + A_{22}\varepsilon_2^0 + A_{62}\varepsilon_6^0\right];$$
$$\overline{\sigma}_3 = \frac{1}{h}\left[A_{61}\varepsilon_1^0 + A_{62}\varepsilon_2^0 + A_{63}\varepsilon_6^0\right] \quad (34)$$

These last equations are known as the effective or global constitutive relations, where the equivalent modulus of a multidirectional laminate is the arithmetic average of the individual modulus of stiffness outside their axis of symmetry of the plies or ply groups. The units of the *Qs* are *Pa (or N / m2)* and the *As* are in $Pa \cdot m$ (or N / m). A stress resultant (*N's*) can be defined, with units of force per unit of length or force per unit of thickness h. This is:

$$N_1 = h\overline{\sigma}_1; N_2 = h\overline{\sigma}_2; N_6 = h\overline{\sigma}_6 \quad (35)$$

Or, equivalently:

$$N_1 = A_{11}\varepsilon_1^0 + A_{21}\varepsilon_2^0 + A_{61}\varepsilon_6^0; N_2 = A_{21}\varepsilon_1^0 + A_{22}\varepsilon_2^0 + A_{63}\varepsilon_6^0;$$
$$N_3 = A_{61}\varepsilon_1^0 + A_{62}\varepsilon_2^0 + A_{66}\varepsilon_6^0 \quad (36)$$

These equations relate the resultant stresses to the strains. To know the stress in each ply or ply group from the strains and global stresses, it is shown schematically in **Figure 14**. First, determine the average strains in terms of the average stresses,

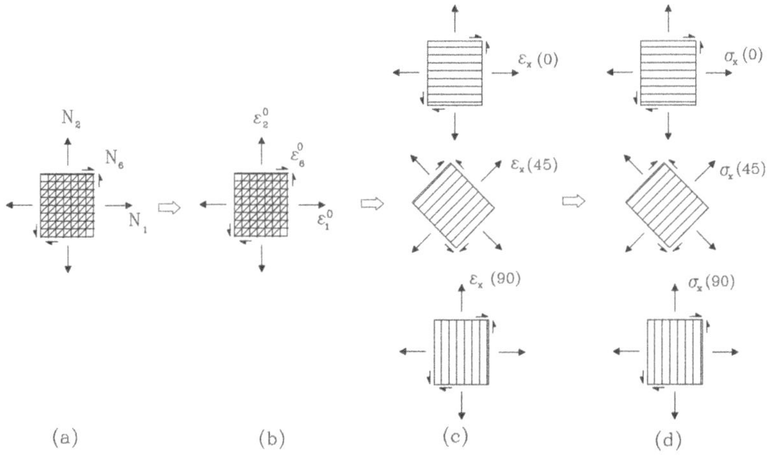

Figure 14.
Process to obtain the state of stress in each ply group from the average state of stress, determining the average strains in terms of the average stresses, (a) to (b), getting the strains for the symmetry axes (system xy), configuration (b) to (c) and finally, determine the state of stress of each ply or plies group, (c) to (d).

for a reference system 12 and apply an effective or global constitutive relations, configuration (a) to (b). Second, getting the strains for the symmetry axes (system xy) of each orthotropic plies or orthotropic ply groups with different orientation, configuration (b) to (c). Finally, apply the constitutive relations to determine the state of stress of each ply or ply group.

Equivalent modulus can also be expressed in terms of the multi-angle equations, that is:

$$
\begin{aligned}
A_{11} &= \int_{-\frac{h}{2}}^{\frac{h}{2}} Q_{11} dz = \int_{-\frac{h}{2}}^{\frac{h}{2}} (U_1 + U_2 \cos 2\theta + U_3 \cos 4\theta) dz \\
&= U_1 \int_{-\frac{h}{2}}^{\frac{h}{2}} dz + U_2 \int_{-\frac{h}{2}}^{\frac{h}{2}} \cos 2\theta dz + U_3 \int_{-\frac{h}{2}}^{\frac{h}{2}} \cos 4\theta dz \\
A_{21} &= \int_{-\frac{h}{2}}^{\frac{h}{2}} Q_{21} dz = \int_{-\frac{h}{2}}^{\frac{h}{2}} (U_4 - U_3 \cos 4\theta) dz = U_4 \int_{-\frac{h}{2}}^{\frac{h}{2}} dz - U_3 \int_{-\frac{h}{2}}^{\frac{h}{2}} \cos 4\theta dz \\
A_{22} &= \int_{-\frac{h}{2}}^{\frac{h}{2}} Q_{22} dz = \int_{-\frac{h}{2}}^{\frac{h}{2}} (U_1 - U_2 \cos 2\theta + U_3 \cos 4\theta) dz \\
&= U_1 \int_{-\frac{h}{2}}^{\frac{h}{2}} dz - U_2 \int_{-\frac{h}{2}}^{\frac{h}{2}} \cos 2\theta dz + U_3 \int_{-\frac{h}{2}}^{\frac{h}{2}} \cos 4\theta dz \\
A_{61} &= \int_{-\frac{h}{2}}^{\frac{h}{2}} Q_{31} dz = \frac{1}{2} U_2 \int_{-\frac{h}{2}}^{\frac{h}{2}} sen 2\theta dz + U_3 sen 4\theta) dz = \frac{1}{2} U_2 \int_{-\frac{h}{2}}^{\frac{h}{2}} sen 2\theta + U_3 \int_{-\frac{h}{2}}^{\frac{h}{2}} sen 4\theta dz \\
A_{62} &= \int_{-\frac{h}{2}}^{\frac{h}{2}} Q_{32} dz = \int_{-\frac{h}{2}}^{\frac{h}{2}} \left(\frac{1}{2} U_2 sen 2\theta - U_3 sen 4\theta \right) dz = \frac{1}{2} U_2 \int_{-\frac{h}{2}}^{\frac{h}{2}} sen 2\theta - U_3 \int_{-\frac{h}{2}}^{\frac{h}{2}} sen 4\theta dz \\
A_{66} &= \int_{-\frac{h}{2}}^{\frac{h}{2}} Q_{33} dz = \int_{-\frac{h}{2}}^{\frac{h}{2}} (U_5 - U_3 \cos 4\theta) dz = U_5 \int_{-\frac{h}{2}}^{\frac{h}{2}} dz - U_3 \int_{-\frac{h}{2}}^{\frac{h}{2}} \cos 4\theta dz
\end{aligned}
\tag{37}
$$

As the Us have no variation with respect to the z axis, they are considered constant. If:

$$
V_1 = \int_{-\frac{h}{2}}^{\frac{h}{2}} \cos 2\theta dz; V_2 = \int_{-\frac{h}{2}}^{\frac{h}{2}} \cos 4\theta dz; V_3 = \int_{-\frac{h}{2}}^{\frac{h}{2}} sen 2\theta dz; V_4 = \int_{-\frac{h}{2}}^{\frac{h}{2}} sen 4\theta dz \tag{38}
$$

Then:

$$
\begin{aligned}
&A_{11} = U_1 h + U_2 V_1 + U_3 V_2; A_{21} = U_4 h - U_3 V_2; \\
&A_{22} = U_1 h - U_2 V_1 + U_3 V_2; A_{61} = \frac{1}{2} U_2 V_3 + U_3 V_4 \\
&A_{62} = \frac{1}{2} U_2 V_3 - U_3 V_4; A_{66} = U_5 h - U_3 V_2
\end{aligned}
\tag{39}
$$

The V values now depend on the orientation of the plies or ply groups in the multidirectional laminate (see **Figure 15**). When normalizing the equations' Vs in terms of the thickness of the laminate so that the values are dimensionless, the following expressions are obtained:

$$V_1^* = \frac{V_1}{h} = \frac{1}{h}\int_{-\frac{h}{2}}^{\frac{h}{2}} \cos 2\theta dz; V_2^* = \frac{V_2}{h} = \frac{1}{h}\int_{-\frac{h}{2}}^{\frac{h}{2}} \cos 4\theta dz;$$
$$V_3^* = \frac{V_3}{h} = \frac{1}{h}\int_{-\frac{h}{2}}^{\frac{h}{2}} sen 2\theta dz; V_4^* = \frac{V_4}{h} = \frac{1}{h}\int_{-\frac{h}{2}}^{\frac{h}{2}} sen 4\theta dz \tag{40}$$

If the plies or ply groups that make up the laminate have the same orientation and the same material, the expressions described above take the following form:

$$V_1^* = \frac{1}{h}\sum_{i=1}^{h} \cos 2\theta_i [z_i - z_{i-1}] = \frac{1}{h}\sum_{i=1}^{h} \cos 2\theta_i h_i;$$
$$V_2^* = \frac{1}{h}\sum_{i=1}^{h} \cos 4\theta_i [z_i - z_{i-1}] = \frac{1}{h}\sum_{i=1}^{h} \cos 4h_i$$
$$V_3^* = \frac{1}{h}\sum_{i=1}^{h} sen 2\theta_i [z_i - z_{i-1}] = \frac{1}{h}\sum_{i=1}^{h} sen 2\theta_i h_i;$$
$$V_4^* = \frac{1}{h}\sum_{i=1}^{h} sen 4\theta_i [z_i - z_{i-1}] = \frac{1}{h}\sum_{i=1}^{h} sen 4\theta_i h_i \tag{41}$$

Where h_i is the thickness of i-th ply group and starts from $\frac{h}{2}$, as in **Figure 15**.

The volumetric fraction can be expressed as follows: $v_i = \frac{h_i}{h}$ and if each i represents an orientation, then Equations (Eq. (41)) are as follows:

$$V_1^* = \sum_{i=1}^{h} \cos 2\theta_i v_i = v_1 \cos 2\theta_1 + v_2 \cos 2\theta_2 + v_3 \cos 2\theta_3 + \dots$$
$$V_2^* = \sum_{i=1}^{h} \cos 4\theta_i v_i = v_1 \cos 4\theta_1 + v_2 \cos 4\theta_2 + v_3 \cos 4\theta_3 + \dots$$
$$V_3^* = \sum_{i=1}^{h} sen 2\theta_i v_i = v_1 sen 2\theta_1 + v_2 sen 2\theta_2 + v_3 sen \theta_3 + \dots$$
$$V_4^* = \sum_{i=1}^{h} sen 4\theta_i v_i = v_1 sen 4\theta_1 + v_2 sen 4\theta_2 + v_3 sen 4\theta_3 + \dots \tag{42}$$

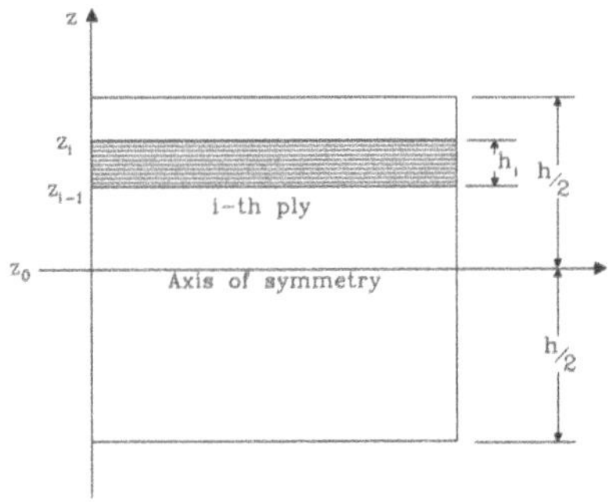

Figure 15.
Graphic representation of n ply groups in a multidirectional symmetric laminate.

The sum of the volume fractions fulfills the condition $v_1 + v_2 + v_3 + \dots = 1$ and the limits of $V^*{}'s$ are $-1 \leq V^*{}'s \geq 1$. By considering the above, the effective or global mechanical properties can be determined, if the orientation and fraction volume of each ply group is known.

5. Theoretical approach to obtain the constitutive models

This section presents the algebraic development to generate the constitutive mathematical model of a laminate composite material taking into account the properties of the constituent orthotropic plies. The laminate to be modeled is made up of longitudinal plies of fiberglass reinforced with epoxy resin, oriented orthogonally. The mechanical engineering properties that characterize an orthotropic laminate are five: 1) Two Young's moduli (*Ex* and *Ey*, two Poisson's ratios ν_{xy} and ν_{yx}, and 2) a shear modulus (*E6 or Gxy*). **Figure 16** shows an orthotropic lamina.

The constitutive relations for a lamina in terms of the engineering constants of each of the layers are obtained by considering equations (Eq. (16)) and (Eq. (27)), that is:

$$U_1 = \frac{1}{8}\left[\frac{3E_x}{1-\nu_x\nu_y} + \frac{3E_y}{1-\nu_x\nu_y} + \frac{2\nu_y E_x}{1-\nu_x\nu_y} + 4E_{ss}\right] = \frac{1}{8}\left[\frac{3E_x + 3E_y + 2\nu_y E_x}{1-\nu_x\nu_y} + 4E_{ss}\right];$$

$$U_2 = \frac{1}{2}\left[\frac{E_x - E_y}{1-\nu_x\nu_y}\right]; U_3 = \frac{1}{8}\left[\frac{E_y + E_x - 2\nu_y E_x}{1-\nu_x\nu_y} - 4E_{ss}\right];$$

$$U_4 = \frac{1}{8}\left[\frac{E_y + E_x + 6\nu_y E_x}{1-\nu_x\nu_y} - 4E_{ss}\right]; U_5 = \frac{1}{8}\left[\frac{E_y + E_x - 2\nu_y E_x}{1-\nu_x\nu_y} + 4E_{ss}\right] \tag{43}$$

By considering equations (Eq. (40)) and (Eq. (42)), the relations for any multidirectional lamina are obtained. This is:

$$\begin{aligned} V_1 &= (\nu_1 cos\ 2\theta_1 + \nu_2 cos\ 2\theta_2 + \nu_3 cos\ 2\theta_3 + \dots)h; \\ V_2 &= (\nu_1 cos\ 4\theta_1 + \nu_2 cos\ 4\theta_2 + \nu_3 cos\ 4\theta_3 + \dots)h; \\ V_3 &= (\nu_1 sen\ 2\theta_1 + \nu_2 sen\ 2\theta_2 + \nu_3 sen\ 2\theta_3 + \dots)h; \\ V_4 &= (\nu_1 sen\ 4\theta_1 + \nu_2 sen\ 4\theta_2 + \nu_3 sen\ 4\theta_3 + \dots)h \end{aligned} \tag{44}$$

From the relations of (Eq. (43)), the following expressions are obtained:

$$\begin{aligned} &A_{11} = U_1 h + U_2 V_1 + U_3 V_2; A_{21} = U_4 h - U_3 V_2; A_{22} = U_1 h - U_2 V_1 + U_3 V_2 \\ &A_{61} = \frac{1}{2} U_2 V_3 + U_3 V_4; A_{62} = \frac{1}{2} U_2 V_3 - U_3 V_4; A_{66} = U_5 h - U_3 V_2 \end{aligned} \tag{45}$$

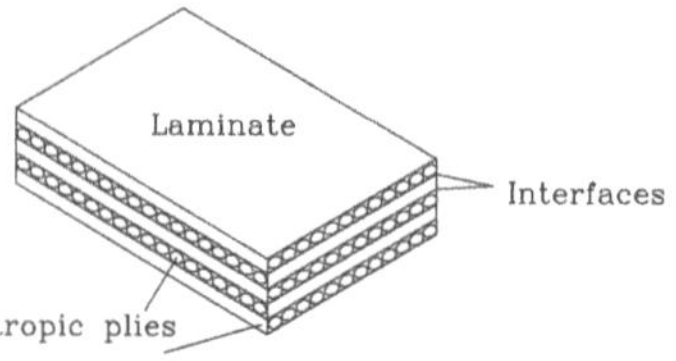

Figure 16.
Orthotropic laminate.

The average values of the modules for "*n*" plies or "*n*" ply groups are obtained by making equations (Eq. (45)) explicit. This is:

$$
\begin{aligned}
A_{11} &= \frac{1}{8}\left[\frac{3E_x + 3E_y + 2\nu_y E_x}{1-\nu_x\nu_y} + 4E_{ss}\right]h \\
&+ \frac{1}{2}\left[\frac{E_x - E_y}{1-\nu_x\nu_y}\right](\nu_1 \cos\ 2\theta_1 + \nu_2 \cos\ 2\theta_2 + \nu_3 \cos\ 2\theta_3 + \ldots)h \\
&+ \frac{1}{8}\left[\frac{E_y + E_x - 2\nu_y E_x}{1-\nu_x\nu_y} - 4E_{ss}\right](\nu_1 \cos\ 4\theta_1 + \nu_2 \cos\ 4\theta_2 + \nu_3 \cos\ 4\theta_3 + \ldots)h \\
A_{21} &= \frac{1}{8}\left[\frac{E_y + E_x + 6\nu_y E_x}{1-\nu_x\nu_y} - 4E_{ss}\right]h \\
&- \frac{1}{8}\left[\frac{E_y + E_x - 2\nu_y E_x}{1-\nu_x\nu_y} - 4E_{ss}\right](\nu_1 \cos\ 4\theta_1 + \nu_2 \cos\ 4\theta_2 + \nu_3 \cos\ 4\theta_3 + \ldots)h \\
A_{22} &= \frac{1}{8}\left[\frac{3E_x + 3E_y + 2\nu_y E_x}{1-\nu_x\nu_y} + 4E_{ss}\right]h \\
&- \frac{1}{2}\left[\frac{E_x - E_y}{1-\nu_x\nu_y}\right](\nu_1 \cos\ 2\theta_1 + \nu_2 \cos\ 2\theta_2 + \nu_3 \cos\ 2\theta_3 + \ldots)h \\
&+ \frac{1}{8}\left[\frac{E_y + E_x - 2\nu_y E_x}{1-\nu_x\nu_y} - 4E_{ss}\right](\nu_1 \cos\ 4\theta_1 + \nu_2 \cos\ 4\theta_2 + \nu_3 \cos\ 4\theta_3 + \ldots)h \\
A_{61} &= \frac{1}{4}\left[\frac{E_x - E_y}{1-\nu_x\nu_y}\right](\nu_1 sen\ 2\theta_1 + \nu_2 sen\ 2\theta_2 + \nu_3 sen\ 2\theta_3 + \ldots)h \\
&+ \frac{1}{8}\left[\frac{E_y + E_x - 2\nu_y E_x}{1-\nu_x\nu_y} - 4E_{ss}\right](\nu_1 sen\ 4\theta_1 + \nu_2 sen\ 4\theta_2 + \nu_3 sen\ 4\theta_3 + \ldots)h \\
A_{62} &= \frac{1}{4}\left[\frac{E_x - E_y}{1-\nu_x\nu_y}\right](\nu_1 sen\ 2\theta_1 + \nu_2 sen\ 2\theta_2 + \nu_3 sen\ 2\theta_3 + \ldots)h \\
&- \frac{1}{8}\left[\frac{E_y + E_x - 2\nu_y E_x}{1-\nu_x\nu_y} - 4E_{ss}\right](\nu_1 sen\ 4\theta_1 + \nu_2 sen\ 4\theta_2 + \nu_3 sen\ 4\theta_3 + \ldots)h \\
A_{66} &= \frac{1}{8}\left[\frac{E_y + E_x - 2\nu_y E_x}{1-\nu_x\nu_y} + 4E_{ss}\right]h \\
&- \frac{1}{8}\left[\frac{E_y + E_x - 2\nu_y E_x}{1-\nu_x\nu_y} - 4E_{ss}\right](\nu_1 \cos\ 4\theta_1 + \nu_2 \cos\ 4\theta_2 + \nu_3 \cos\ 4\theta_3 + \ldots)h
\end{aligned} \tag{46}
$$

By substituting equations (Eq. (46)) in equations (Eq. (36)), the following relations are obtained:

$$
\begin{aligned}
&N_1 = A_{11}\varepsilon_1^0 + A_{21}\varepsilon_2^0 + A_{61}\varepsilon_6^0; N_2 = A_{21}\varepsilon_1^0 + A_{22}\varepsilon_2^0 + A_{62}\varepsilon_6^0; \\
&N_6 = A_{61}\varepsilon_1^0 + A_{62}\varepsilon_2^0 + A_{66}\varepsilon_6^0
\end{aligned} \tag{47}
$$

And, when considering the expression (Eq. (35)) the following relations are obtained:

$$\bar{\sigma}_1 = \frac{h}{N_1}; \bar{\sigma}_2 = \frac{h}{N_2}; \bar{\sigma}_6 = \frac{h}{N_6} \tag{48}$$

If the volume fractions and the orientations of each ply group are known, equations (Eq. (48)) represent the constitutive relations for any multidirectional laminate. These equations contain global or effective properties A's and the average stresses and strains.

5.1 Constitutive equations for laminate

The constitutive model for a configuration laminate $\left[0^0/90^0/0^0/90^0/0^0\right]_T$ is obtained from equations (Eq. (47)) and (Eq. (48)) by considering $\theta = 0^0$ and $\theta = 90^0$ with volume fractions $v_1 = \frac{3}{5}$ and $v_2 = \frac{2}{5}$. This is:

$$\bar{\sigma}_1 = \frac{1}{5}\left[\frac{3E_x + 2E_y}{1 - \nu_x\nu_y}\right]\varepsilon_1^0 + \left[\frac{\nu_y E_x}{1 - \nu_x\nu_y}\right]\varepsilon_2^0; \bar{\sigma}_2 = \left[\frac{\nu_y E_x}{1 - \nu_x\nu_y}\right]\varepsilon_1^0 + \frac{1}{5}\left[\frac{2E_x + 3E_y}{1 - \nu_x\nu_y}\right]\varepsilon_2^0; \bar{\sigma}_6 = E_{ss}\varepsilon_6^0 \tag{49}$$

For the case of the laminate $\left[90^0/0^0/90^0/0^0/90^0\right]_T$ the constitutive model is obtained considering $\theta = 90^0$ and $\theta = 0^0$ with volumetric fractions $v_1 = \frac{2}{5}$ y $v_2 = \frac{3}{5}$. This is:

$$\bar{\sigma}_1 = \frac{1}{5}\left[\frac{2E_x + 3E_y}{1 - \nu_x\nu_y}\right]\varepsilon_1^0 + \left[\frac{\nu_y E_x}{1 - \nu_x\nu_y}\right]\varepsilon_2^0; \bar{\sigma}_2 = \left[\frac{\nu_y E_x}{1 - \nu_x\nu_y}\right]\varepsilon_1^0 + \frac{1}{5}\left[\frac{3E_x + 2E_y}{1 - \nu_x\nu_y}\right]\varepsilon_2^0; \bar{\sigma}_6 = E_{ss}\varepsilon_6^0 \tag{50}$$

Finally, the constitutive model for the laminate $\left[0^0/90^0/0^0\right]_T$ is generated considering that $\theta = 0^0$ and $\theta = 90^0$ with volume fractions $v_1 = \frac{2}{5}$ and $v_2 = \frac{3}{5}$. The model is as follows:

$$\bar{\sigma}_1 = \frac{1}{3}\left[\frac{2E_x + E_y}{1 - \nu_x\nu_y}\right]\varepsilon_1^0 + \left[\frac{\nu_y E_x}{1 - \nu_x\nu_y}\right]\varepsilon_2^0; \bar{\sigma}_2 = \left[\frac{\nu_y E_x}{1 - \nu_x\nu_y}\right]\varepsilon_1^0 + \frac{1}{3}\left[\frac{2E_x + E_y}{1 - \nu_x\nu_y}\right]\varepsilon_2^0; \bar{\sigma}_6 = E_{ss}\varepsilon_6^0 \tag{51}$$

6. Experimental obtaining of the elastic engineering properties to ply

In order to obtain the engineering properties of laminated plies, the experimental electrical extensometry method is used, which is supported by Perry [18]. It is worth mentioning that, due to the arrangement of plies in the proposed composite laminate, the number of constitutive analytical equations are three and the number of elastic constants of orthotropic plies are five, so five- and three-ply laminate are used. A CEA-O6-240UZ-120 strain gage is selected for the study (see **Figure 17**).

For the experimental tests, an INSTRON Universal Testing Machine was used. The engineering properties E_x, E_y, ν_x and ν_y were determined. Several tests were carried out and, in the solution, the equations of two models were combined due to the number of constants to be determined.

Figure 18 shows the tests as well as the location of the installed strain gages, two for each deformation in a full Wheatstone bridge configuration to eliminate signals outside the required measurement (deflections, temperature changes, etc.). Tests were performed with laminates $\left[0^0/90^0/0^0/90^0/0^0\right]_T$ and $\left[90^0/0^0/90^0/0^0/90^0\right]_T$ for characterization and $\left[0^0/90^0/0^0\right]_T$ for evaluation of results. The measurement

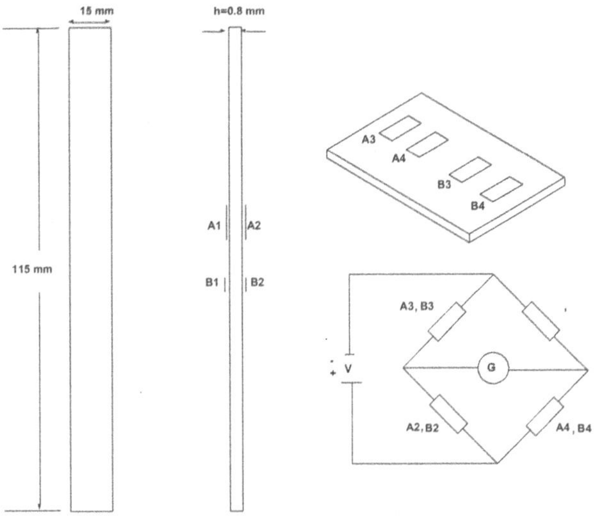

Figure 17.
Installation of strain gages, active and dummy.

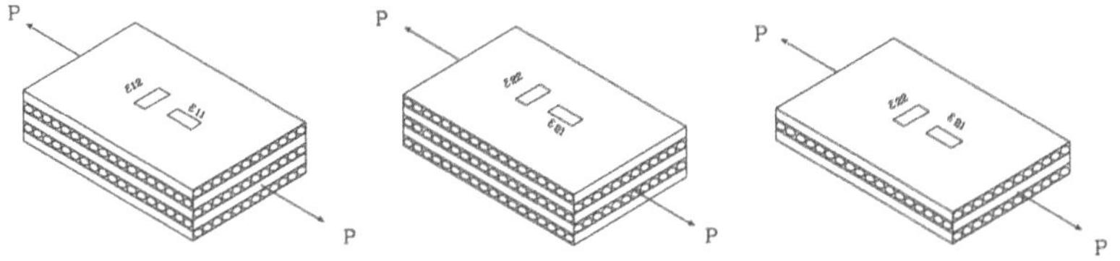

Figure 18.
Tension test system with test specimens 3–2, 2–3 and 2–1 and installation of strain gages.

of the strains was carried out with electrical resistance strain gages on a theoretically homogenized surface, and the values obtained are average and not punctual. From the readings provided by the meter, the effective properties for both a laminate and plies are calculated.

Combining test specimens 3–2 and 2–3 and from equations (Eq. (49)) and (Eq. (50)), the following equations are obtained to determine the constants E_x, E_y, ν_x and ν_y:

$$\begin{aligned} v_x &= \frac{\varepsilon_{12}^0 \varepsilon_{22}^0}{\left(2\varepsilon_{12}^0 \varepsilon_{21}^0 - 3\varepsilon_{11}^0 \varepsilon_{22}^0\right)}; \quad v_y = \frac{\varepsilon_{12}^0 \varepsilon_{22}^0}{\left(2\varepsilon_{11}^0 \varepsilon_{22}^0 - 3\varepsilon_{12}^0 \varepsilon_{21}^0\right)} \\ E_x &= \frac{5\overline{\sigma}_{11} v_x \left(1 - v_x v_y\right)}{3v_x \varepsilon_{11}^0 + 2v_y \varepsilon_{11}^0 + 5v_x v_y \varepsilon_{12}^0}; \quad E_y = \frac{5\overline{\sigma}_{11} v_y \left(1 - v_x v_y\right)}{\left(3v_x \varepsilon_{11}^0 + 2v_y \varepsilon_{11}^0 + 5v_x v_y \varepsilon_{12}^0\right)} \end{aligned} \tag{52}$$

Here, $\overline{\sigma}_{11}$, $\overline{\sigma}_{12}$, ε_{11}^0, ε_{12}^0 are the global average stress and strain values, respectively, in directions 1 and 2 for the laminate $\left[0^0/90^0/0^0/90^0/0^0\right]_T$, and $\overline{\sigma}_{21}$, $\overline{\sigma}_{22}$, ε_{21}^0, ε_{22}^0 are the average stress and strain values in directions 1 and 2 for the laminate $\left[90^0/0^0/9\ 0^0/0^0/90^0\right]_T$. In the tests, $\overline{\sigma}_{12} = \overline{\sigma}_{22} = 0$.

6.1 Analysis of results

The analysis of the data complied with the symmetry of the stiffness and compliance tensor, the identity $\frac{v_x}{E_x} = \frac{v_y}{E_y}$, for calculated values of a ply. For the results of

Engineering Constants Plies	Magnitude
E_x	53.05 GPa
E_y	23.3 GPa
ν_x	0.233
ν_y	0.1

Table 1.
Properties of the average experimental engineering constants.

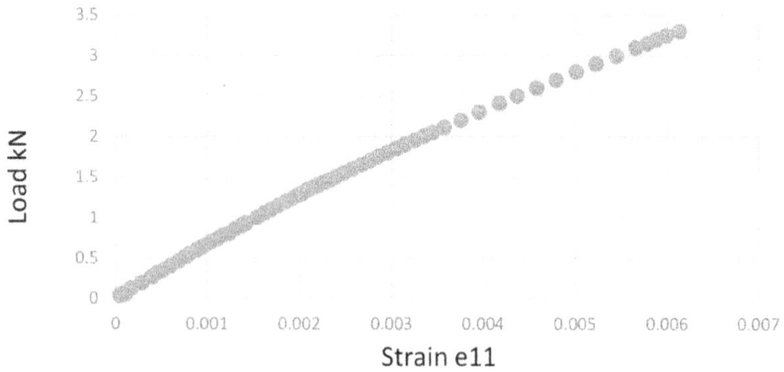

Figure 19.
Graph of laminate 3–2 under tension.

the effective mechanical properties obtained experimentally E_x, E_y, ν_x and ν_y of the plies that make up the laminate composite material, see **Table 1**. **Figure 19** shows a representative test on laminate 3–2.

The consistency of the tests was carried out by comparing the results obtained for tests 3–2 and 2–1, showing very proximate values.

7. Conclusions

In this chapter, a theoretical framework based on the theory of linear elasticity and the classical theory of composite laminates was established for the analysis of axial load problems by analyzing concepts and establishing scopes and restrictions of the models applied by the theories. The necessary bases were established for the general obtention of the composite laminates' constitutive models made up of orthotropic plies or orthotropic ply groups with different orientations. With the established models, it is possible: a) with the known stiffness constants, to calculate the state of plane stresses at a point from the state of deformations and b) with the known conformity constants, to calculate the state of strains from the state of stresses. For ease of use, both the stiffness constants and the compliance constants were made explicit in terms of the engineering constants. This allows an analysis of overall or average stresses in the laminates under axial load. To show the efficiency of the developments presented, the constitutive models of three orthotropic composite laminates were obtained. Furthermore, by performing simple stress tests on the laminates and measuring the state of strain with strain gages, the engineering constants of the plies were determined.

Author details

Mario Acosta Flores[1], Eusebio Jiménez López[2]* and Marta Lilia Eraña Díaz[1]

1 Autonomous University of the State of Morelos, Cuernavaca, México

2 Technological University of Southern Sonora - La Salle University Northwest, Obregón, México

*Address all correspondence to: ejimenezl@msn.com

References

[1] Shashi B. 2021. Fiber reinforced metal matrix composites - a review, Materials Today: Proceedings. 2021; 39: 2214-7853. DOI: 10.1016/j.matpr. 2020.07.423

[2] Gholizadeh S. A review of impact behaviour in composite materials. International Journal of Mechanical and Production Engineering, 2019; 7: 35-46.

[3] Nurazzi M, Khalina A, Sapuan S. M, Dayang L, Rahmah M, and Hanafee Z. A Review: Fibres, Polymer Matrices and Composites, Pertanika J. Sci. & Technol. 2017; 25: 1085 – 1102

[4] Abdellaoui H, Bensalah H, Raji M, Rodrigue D, Bouhfid R. Laminated epoxy biocomposites based on clay and jute fibers. J Bionic Eng. 2017; 14: 379-389. DOI: 10.1016/S1672-6529(16) 60406-7

[5] Pagano J. N, Pipes R B. Some observations on the interlaminar strength of composite laminates. International Journal of Mechanical Sciences. 1973; 15: 679-686. DOI: 10.1016/0020-7403(73)90099-4

[6] Elanchezhian C, Vijaya B, Ramakrishnan G, Rajendrakumar M, Naveenkumare V, and Saravanakumarf M. K. (2018). Review on mechanical properties of natural fiber composites, Materials Today: Proceedings; 2018; 5:1785–1790. DOI: 10.1016/j.matpr.2017.11.276

[7] Sinha L, Mishra S. S, Nayak A. N, and Sahu S. K. (2020). Free vibration characteristics of laminated composite stiffened plates: Experimental and numerical investigation. Composite Structures; 2020; 233, 1111557: DOI: 10.1016/j.compstruct.2019.111557

[8] Sun G, Chen D, Huo X, Zheng G, and Li Q. Experimental and numerical studies on indentation and perforation characteristics of honeycomb sandwich panels. Composite Structures, 2018; 184, 15: 110-124. DOI: 10.1016/j. compstruct.2017.09.025

[9] Acosta M, Eraña M. L, Jiménez E, García G. C, Delfín J. J, and Lucero B. (2021). Analytical method for determining maximum shear stresses in laminated composite metal bars subjected to torsion. J Mech Sci Technol. 2021; 35: 3019–3031. DOI: 10.1007/ s12206-021-0625-x

[10] D'Antino T, and Papanicolaou C. Mechanical characterization of textile reinforced inorganic-matrix composites. Composites Part B; 2017, 127: 78-91. DOI: 10.1016/j.compositesb.2017. 02.034.

[11] Hashem L, Mottar Al M. M, Hussien S, and Abed A. M. Experimental study the mechanical properties of nano composite materials by using multi-metallic nano powder/ epoxy, Materials Today: Proceedings, 2021. DOI: 10.1016/j.matpr.2021.06.395.

[12] Acosta M, Jiménez E, Chávez M, Molina A, and Delfín, J. J. (2019). Experimental method for obtaining the elastic properties of components of a laminated composite, Results in Physics, 2019; 12: 1500-1505. DOI: 10.1016/j. rinp.2019.01.016

[13] Xu Y, Zhu J, Wu Z, Cao Y, Zhao Y, and Zhang W. A review on the design of laminated composite structures: constant and variable stiffness design and topology optimizatio. Advanced Composites and Hybrid Materials, 2018; 1: 460-477. DOI: 10.1007/s42114-018-0032-7

[14] Durelli A, Phillips E, Tsao C. Introduction To The Theoretical and Experimental Analysis of Stress and Strain. 1st ed. New York, USA: McGraw-Hill Book Company, Inc. 1958.

[15] Dally J, Riley F. Experimental Strees Analysis. 2nd ed. New York, USA: McGraw-Hill; 2005.

[16] Tsai SW, Hahn HT. Introduction to Composite Materials. 1st ed. Pennsylvania, USA: Technomic Publishing Co.; 1980.

[17] Jones, Roberts M., Mechanics of Composite Materials, 2nd ed. New York, USA: McGraw-Hill Book Company, Inc. 1999.

[18] Perry C. C. Strain-Gage Reinforcement Effects on Orthotropic Materials, In: Pendleton R.L., Tuttle M. E. (eds) Manual on Experimental Methods for Mechanical Testing of Composites. Springer, Dordrecht. 1989. P. 39-44. DOI:10.1007/978-94-009-1129-1_8

Section 2

Characterization

Chapter 5

Temperature Dependence of the Stress Due to Additives in KCl Single Crystals

Yohichi Kohzuki

Abstract

The influence of the state of additive cations on the various deformation characteristics was studied for KCl:Sr^{2+} single crystal at room temperature. This result gives the heat treatment suitable for the crystal immediately before deformation tests, such as compression and tension. Four kinds of single crystals (KCl: Mg^{2+}, Ca^{2+}, Sr^{2+} or Ba^{2+}) were plastically deformed by compression at 77 to room temperature. The plasticity of the crystal depends on dislocation motion from a microscopic viewpoint. When a dislocation breaks away from the defect around the additive cation with the help of thermal activation on the slip plane in the crystal, the variation of effective stress with the temperature was investigated by the combination method of strain-rate cycling tests and ultrasonic oscillations. Furthermore, the critical temperature T_c at which the effective stress due to the additives is zero was estimated for each of the crystals. As a result, T_c value tends to be larger with the divalent cationic size.

Keywords: dislocation, divalent cation, effective stress, yield stress, heat treatment

1. Introduction

In alkali halide crystals doped with divalent cations (divalent impurities), the additive cations are expected to be paired with vacancies to conserve the electrical neutrality. They are often formed a divalent impurity-vacancy (I-V) dipole for the impure crystals quenched from a high temperature. Then, the asymmetrical distortions are produced around the I-V dipoles. Mobile dislocations on a slip plane interact strongly only with these defects lying within one atom spacing of the glide plane [1]. The solution hardening is named "rapid hardening," which can be distinguished from "gradual hardening" due to the defects of cubic symmetry around the monovalent dopants in the crystals [2–4]. Effects of different defects on the hardness of some alkali halide crystals are listed in **Table 1**. The effects are expressed as an increase in flow stress per square root of concentration of point defects (i.e., $\Delta\tau$ / $\Delta c^{1/2}$) in terms of the shear modulus, μ. Despite the same matrix (see NaCl in **Table 1**), the hardening due to substitutional divalent additions is much larger than the case of monovalent ones. It has been well known for many years that aliovalent impurities (aliovalent cations) are a much more potent source of solution strengthening in ionic crystals than isovalent cations [5].

Crystal	Different types of point defects	$\Delta\tau/\Delta c^{1/2}$ ($\times 2\sqrt{c}$) [a]
NaCl	Monovalent substitutional impurities	$< \mu/30$
KCl	F-centers, additive coloring	$\mu/2.5$
NaCl	Divalent substitutional impurities	2μ
KCl (irradiated)	Interstitial chlorine	18μ
LiF	Divalent impurity clusters	6μ
LiF	I-V dipoles (at 77 K)	10μ
LiF (irradiated)	V_K-centers (at 77 K)	25μ
LiF (irradiated)	Interstitial fluorine	5μ

[a] *Δc represents the increment of the concentration of point defects and μ is the shear modulus. The measurements were made at room temperature unless otherwise noted. $\Delta\tau/\Delta c$ in Refs. [2–4] is replaced by $\Delta\tau/\Delta c^{1/2}$.*

Table 1.
Hardening due to various defects in alkali halide crystals. Defects concentration is below 10^{-4} [2–4].

In view of the different types of atomic defects, solution hardening may be divided into two classes: rapid hardening and gradual hardening by Fleischer and Hibbard [3] and Johnston et al. [4]. Roughly speaking, the value of $\Delta\tau/\Delta c^{1/2}$ for rapid hardening is over several ten times as large as that for the gradual hardening as shown in **Table 1**.

It is well known that various characteristics of deformation are influenced by the state of impurities in a crystal. The following is concerned with it and presents the heat-treatment condition suitable for the deformation tests such as compression and tension for $KCl{:}Sr^{2+}$ crystals. Furthermore, the influence of divalent cationic size on the deformation characteristics is reported analyzing the data obtained by the original method (strain-rate cycling tests associated with ultrasonic oscillation), which can separate the effective stress due to weak obstacles such as additive ions from that due to dislocation cuttings.

2. Experimental procedure

The initial dislocation density, the dielectric loss peak due to the I-V dipoles, and yield stress were measured for $KCl{:}Sr^{2+}$ crystal as explained in Sections 2.2 to 2.4.

2.1 Preparation of specimens

KCl doped with $SrCl_2$ was grown from the melt of reagent-grade powders by the Kyropoulos method in air. The specimens, which were cloven out of single-crystalline ingots to the size of $5 \times 5 \times 15$ mm^3, were kept immediately at 973 K for 24 h, followed by cooling to room temperature at a rate of 40 Kh^{-1}. This treatment is because the density of dislocations is reduced as much as possible. Owing to the gradual cooling, the additive ions (Sr^{2+}) are expected to aggregate in the crystal. Accordingly, the specimens were further kept at 373 to 873 K for 30 min, followed by quenching to room temperature to disperse the additive ions (Sr^{2+}) into them.

2.2 Initial dislocation density (ρ)

Using an etch pits technique, the initial density (ρ) of dislocations in $KCl{:}Sr^{2+}$ (0.3 mol.% in the melt) was detected with a corrosive liquid (saturated solution of $PbCl_2$ + ethyl alcohol added two drops of water). The etching was made at room temperature for 30 min. The measurement of dislocation density in a crystal was carried out by the etch pit technique.

2.3 Dielectric loss factor (tan δ)

The dielectric loss factor tan δ as a function of frequency was measured for KCl: Sr^{2+} (1.0 mol.% in the melt) in a thermostatic bath at 300 to 873 K by using an Andoh electricity TR-10C model.

2.4 Yield stress (τ_y)

The values of yield stress τ_y were obtained at room temperature for KCl:Sr^{2+} (1.0 mol.% in the melt) compressed along the <100 > axis at the crosshead speed of 20 μm min^{-1}. The τ_y values were determined by the intersection of the tangent to the easy glide region in the stress-strain curve and the straight line extrapolated from the elastic region of the curve.

2.5 Combination method of strain-rate cycling tests and ultrasonic oscillation

Four kinds of single crystals (KCl:Mg^{2+} (0.035 mol.% in the melt), Ca^{2+} (0.050, 0.065 mol.% in the melt), Sr^{2+} (0.035, 0.050, 0.065 mol.% in the melt) or Ba^{2+} (0.050, 0.065 mol.% in the melt)) were prepared by cleaving the single crystalline ingots to the size of 5 × 5 × 15 mm^3. The test pieces were kept immediately below the melting point (1043 K) for 24 h and were gradually cooled to room temperature at a rate of 40 Kh^{-1}. Further, they were held at 673 K for 30 min and were rapidly cooled by water-quenching immediately before the following tests.

The heat-treated test pieces were compressed along the <100> axis at 77 K to the room temperature and the ultrasonic oscillatory stress (τ_v) was intermittently superimposed in the same direction as the compression. The strain-rate cycling test off or on the ultrasonic oscillation (20 kHz) is illustrated in **Figure 1**. Superposing oscillatory stress, a stress drop ($\Delta\tau$) is caused during plastic deformation. The strain-rate cycling between strain-rates of $\dot{\varepsilon}_1$ (2.2×10^{-5} s^{-1}) and $\dot{\varepsilon}_2$ (1.1×10^{-4} s^{-1}) was undertaken keeping the stress amplitude of τ_v constant. This led to the increase $\Delta\tau$' in stress due to the strain-rate cycling. The strain-rate sensitivity ($\Delta\tau'/\Delta\ln\dot{\varepsilon}$) of the flow stress, which is derived from $\Delta\tau'/1.609$, was used as a measurement of the strain-rate sensitivity.

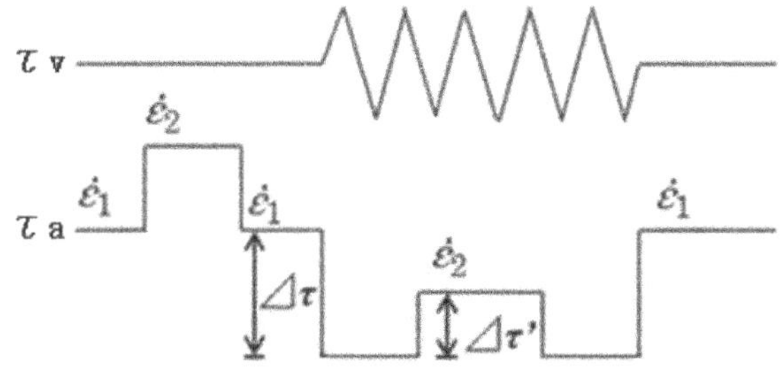

Figure 1.
Change in applied shear stress (τ_a) for the strain-rate cycling tests between the two strain rates, $\dot{\varepsilon}_1$ (2.2×10^{-5} s^{-1}) and $\dot{\varepsilon}_2$ (1.1×10^{-4} s^{-1}), off or on the ultrasonic oscillatory stress (τ_v) due to the oscillation (20 kHz).

3. Results and discussion

3.1 Deformation characteristics influenced by different heat treatments for KCl:Sr^{2+} crystals

Figure 2 shows the optical micrograph of the etch pits for KCl:Sr^{2+} (0.05 mol.% in the melt) at room temperature after the annealing at 973 K for 24 h. The position of the dislocation after the annealed treatment is marked by a pyramidal pit, and the position where a dislocation slipped out of the crystal after the treatment is marked by a flat-bottom pit.

Although, it is difficult to resolve the individual etch pits for high dislocation density, the dislocation density on a (100) plane is found to be 1.27×10^4 cm^{-2} from this micrograph for the annealed specimen (i.e., KCl:Sr^{2+} (0.05 mol.% in the melt)).

The height of the loss peak is related to the concentration of the isolated I-V dipole (see Eq. (1)). The details are explained about KCl:Sr^{2+} (0.05 mol.% in the melt) below.

Dielectric absorption of an I-V dipole causes a peak on the relative curve of tan δ -frequency. The Debye peak height is proportional to the concentration of I-V dipoles as expressed by the following Eq. (1) [6].

$$\tan\delta = \frac{2\pi e^2 c}{3\varepsilon' akT}, \text{ (maximum)} \tag{1}$$

where e is the elementary electric charge, c is the concentration of I-V dipoles, ε' is the dielectric constant in the matrix, a is the lattice constant, k is the Boltzmann constant, and T is the absolute temperature. **Figure 3** shows the tan δ-frequency curves for KCl:Sr^{2+} at 393 K. The solid and dotted curves correspond to the quenched KCl:Sr^{2+} (0.05 mol.% in the melt). That is to say, the crystals were held

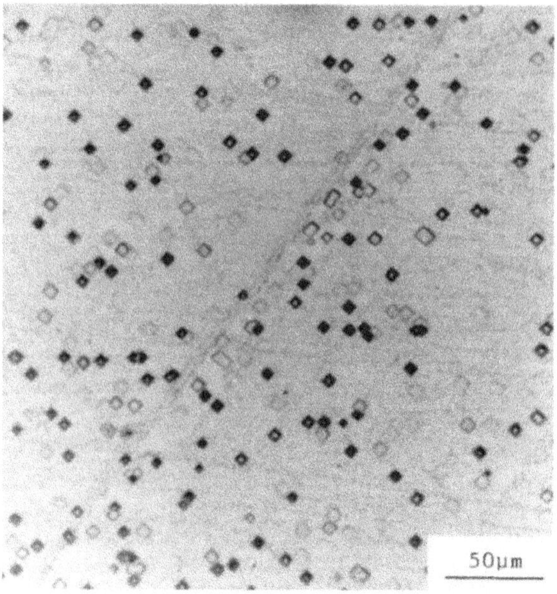

Figure 2.
Dislocations on a (100) plane for KCl:Sr^{2+} (0.05 mol.%) after the heat treatment at 973 K for 24 h.

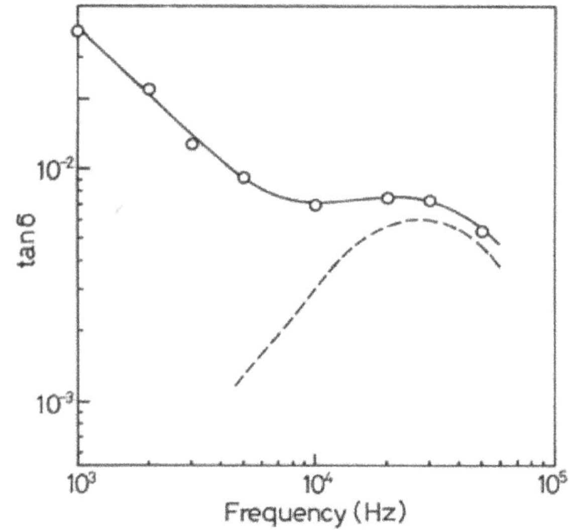

Figure 3.
Dielectric loss in KCl:Sr^{2+}(0.05 mol.% in the melt) at 393 K. Dotted line (— — - -) shows the losses coming from the I-V dipoles.

within 673 K for 30 min, followed by quenching to room temperature. The dotted line shows Debye peak obtained by subtracting the d.c. part which is obtained by extrapolating the linear part of the solid curve in the low-frequency region to the high-frequency region. By introducing the peak height of the dotted curve into Eq. (1), the concentration of the isolated I-V dipoles was determined to be 98.3 ppm for the quenched crystal by dielectric loss measurement.

As mentioned above, the dielectric loss factor tan δ is proportional to the concentration of the isolated I-V dipoles at a given temperature.

Figure 4 shows the variations in the initial dislocation density (ρ), the dielectric loss peak due to the I-V dipoles (tan δ), and yield stress (τ_y) as against the temperature quenched KCl:Sr^{2+} (0.3 and 1.0 mol.% in the melt) single crystals [7]. The ρ value is about $5 \pm 1 \times 10^4$cm^{-2} independently of quenching temperature below 673 K, but it remarkably increases above 673 K. The τ_y value also remarkably increases for the crystals quenched at the temperature above 673 K as the variation in dislocation density. And then it becomes a constant value 29 MPa above 723 K. While the tan δ value does not vary and is almost constant by quenching from the temperature below 573 K or above 673 K. Its value becomes 0.3×10^{-2} up to 0.9×10^{-2} between the two quenching temperatures (i.e., 573 K and 673 K). The variation in tan δ value is similar to it in the yield stress within the temperature.

The difference in dislocation density is slight and the tan δ obviously becomes larger with a higher quenching temperature between 573 and 673 K as shown in **Figure 4**. The concentration of isolated I-V dipoles, which is proportional to the tan δ (see Eq. (1)), affects the yield stress, as reported in the papers [8–11]. Therefore, the specimens are determined to be quenched from 673 K to room temperature immediately before deformation tests such as compression in this chapter.

3.2 Temperature dependence of τ_{p1}, τ_{p2}, and yield stress (τ_y)

The variation of the strain-rate sensitivity and the stress decrement with the shear strain is shown in **Figure 5** for KCl:Sr^{2+} (0.050 mol.% in melt) single crystals at 200 K. $\Delta\tau'/\Delta\ln\dot{\varepsilon}$ tends to increase with shear strain and decrease with stress amplitude in

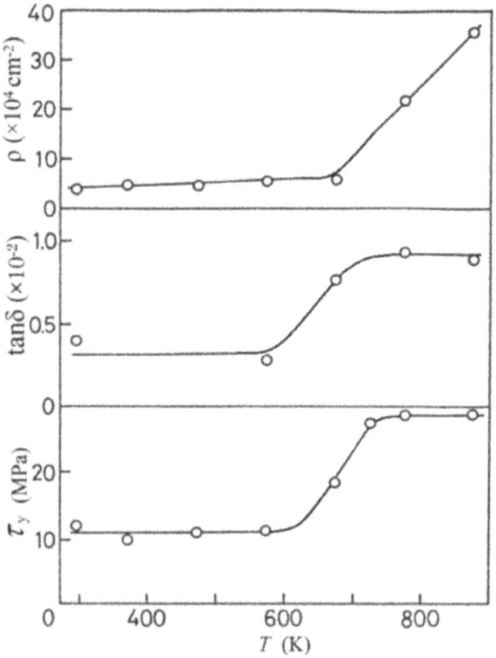

Figure 4.
Quenching temperature dependence of initial dislocation density (ρ), the dielectric loss peaks due to the I-V dipoles ($\tan\delta$), and yield stress (τ_y) for KCl:Sr^{2+} crystals (reproduced from Ref. [7]).

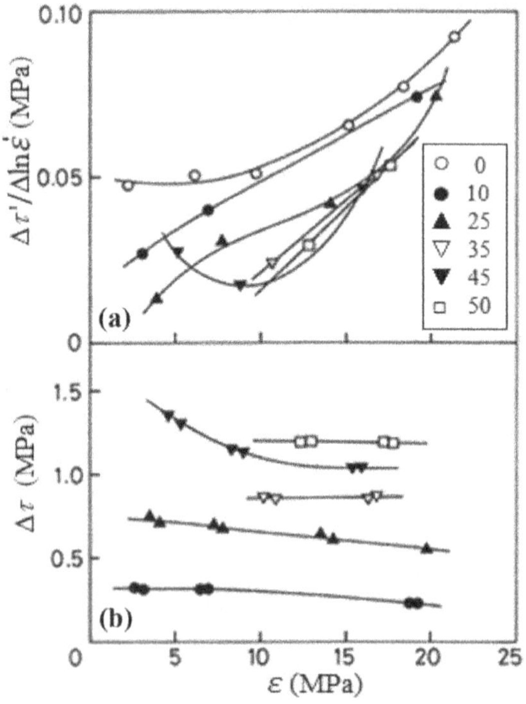

Figure 5.
Shear strain (ε) dependence of (a) the strain-rate sensitivity ($\Delta\tau'/\Delta\ln\dot{\varepsilon}$) and (b) the stress decrement ($\Delta\tau$) for KCl:Sr^{2+} (0.050 mol.%) at 200 K. τ_v (arb. units): (○) 0, (●) 10, (▲) 25, (▽) 35, (▼) 45, and (□) 50 (reproduced from Ref. [12] with permission from the publisher).

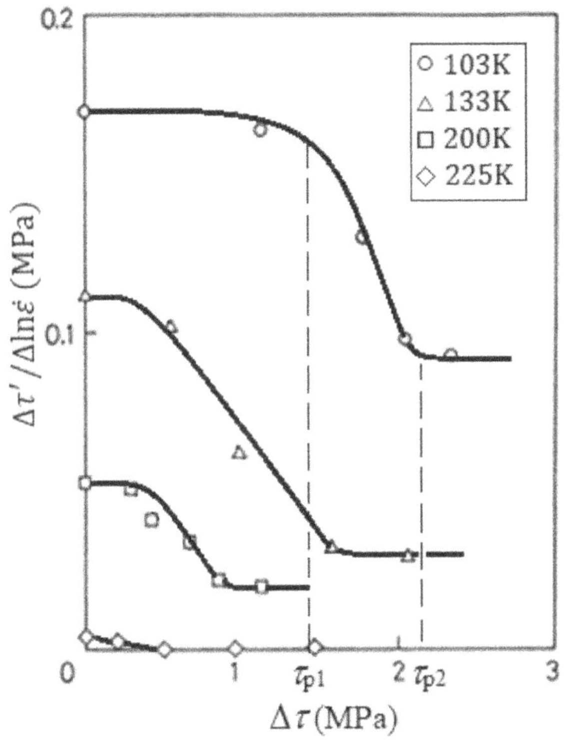

Figure 6.
Strain-rate sensitivity ($\Delta\tau'/\Delta\ln\dot{\varepsilon}$) vs. the stress decrement ($\Delta\tau$) at strain of 10% for KCl:Sr^{2+} (0.050 mol.%) at temperatures of (○) 103 K, (Δ) 133 K, (□) 200 K, (◇) 225 K *(reproduced from Ref. [13] with permission from the publisher).*

Figure 5(a). $\Delta\tau$ does not change significantly with shear strain but increases with stress amplitude at a given temperature and shear strain in **Figure 5(b)**.

$\Delta\tau'/\Delta\ln\dot{\varepsilon}$ vs. $\Delta\tau$ curve is further obtained from **Figure 5** at a given strain, which provides the relative curve for a fixed internal structure of the crystal and is shown by open squares in **Figure 6** for KCl:Sr^{2+} (0.050 mol.% in melt) crystal at the shear strain of 10%. The details were described in the review article [14].

Relation between the strain-rate sensitivity and the stress decrement for KCl:Sr^{2+} (0.050 mol.% in melt) at the shear strain of 10% is shown by open symbols in **Figure 6**. The relative curve has a stair like shape. **Figure 6** shows the influence of temperature on the relation between the strain-rate sensitivity, $\Delta\tau'/\Delta\ln\dot{\varepsilon}$, and the stress decrement, $\Delta\tau$, for KCl single crystals doped with Sr^{2+} as weak obstacles. As the temperature is high, the $\Delta\tau$ value at first bending point, τ_{p1}, shifts in the direction of low stress decrement and does not appear up to 225 K. The first plateau region indicates that the average length of the dislocation segment remains constant in that region. This is because the strain-rate sensitivity of effective stress (τ^*) due to impurities is inversely proportional to the average length of the dislocation segment. That is to say, it is given by

$$\left(\frac{\Delta\tau^*}{\Delta\ln\dot{\varepsilon}}\right)_T = \frac{kT}{bLd} \quad (2)$$

where b is the magnitude of the Burgers vector, L is the average length of dislocation segments, and d is the activation distance. Therefore, the application of

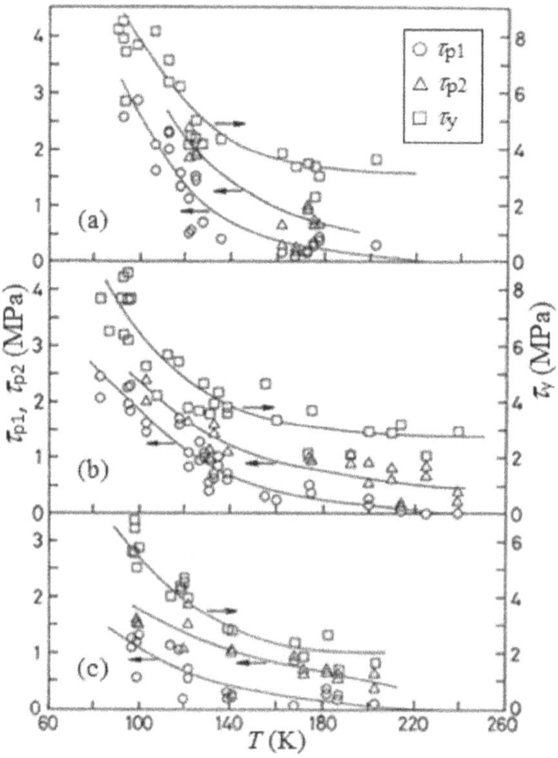

Figure 7.
Temperature dependence of (○) τ_{p1}, (Δ) τ_{p2}, and (□) τ_y for KCl:Sr^{2+} ((a) 0.065, (b) 0.050, (c) 0.035 mol.% in the melt) *(reproduced from Ref. [12]).*

oscillation with low-stress amplitude cannot influence the average length of the dislocation segment at low temperature, but even a low-stress amplitude can do so at a temperature of 225 K. Such a phenomenon was also observed for the other specimens: KCl doped with Mg^{2+}, Ca^{2+} or Ba^{2+} separately.

Figure 7(a)–(c) shows the dependence of τ_{p1}, τ_{p2}, and yield stress (τ_y) on temperature for KCl:Sr^{2+} ((a) 0.065, (b) 0.050 and (c) 0.035 mol.%, respectively) crystals. τ_{p2} is the $\Delta\tau$ value at the second bending point on the plots of $\Delta\tau$ vs. $(\Delta\tau'/\Delta\ln\dot{\varepsilon})$. It is clear from the figure that both τ_{p1} and τ_{p2} tend to increase with decreasing temperature as well as τ_y for the three crystals and the τ_y curve seems to approach a constant stress at high temperature. Two values of τ_{p1} and τ_{p2} increase with increasing Sr^{2+} concentration at a given temperature as shown in the figure. Similar results as the case of KCl:Sr^{2+} are also observed for the other crystals (i.e., KCl: Mg^{2+}, Ca^{2+} or Ba^{2+}).

3.3 Critical temperature (T_c)

Figure 8(a)–(d) shows the dependence of τ_{p1} on the temperature for KCl doped with Mg^{2+} (0.035 mol.% in melt), Ca^{2+} (0.065 mol.% in melt), Sr^{2+} (0.050 mol.% in melt) or Ba^{2+} (0.065 mol.% in melt). τ_{p1} is considered the effective stress due to the weak obstacles (Mg^{2+}, Ca^{2+}, Sr^{2+} or Ba^{2+} ions in this Section 3.2) on the mobile dislocation during plastic deformation [13]. τ_{p1} decreases with increasing temperature for the four kinds of crystals in the figure. The critical temperature (T_c) at which τ_{p1} is zero can be determined from the

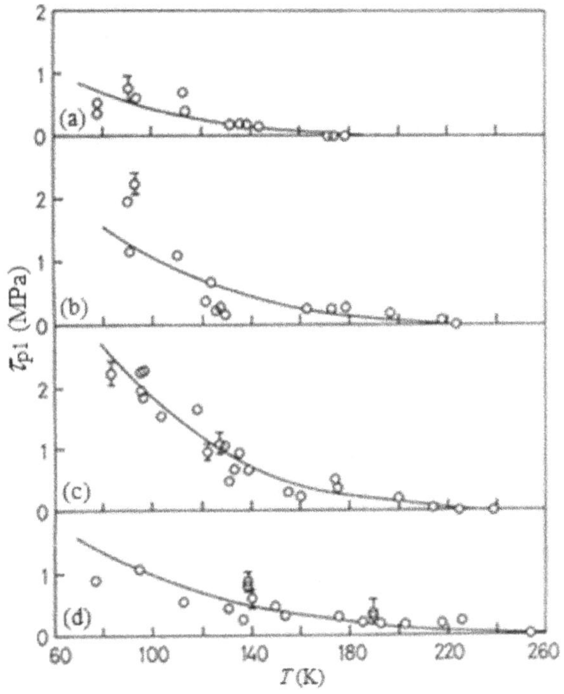

Figure 8.
Temperature dependence of τ_{p1} for various crystals: (a) KCl:Mg^{2+} (0.035 mol.% in the melt), (b) KCl:Ca^{2+} (0.065 mol.% in the melt), (c) KCl:Sr^{2+} (0.050 mol.% in the melt), and (d) KCl:Ba^{2+} (0.065 mol.% in the melt). (Reproduced from Ref. [13] with permission from the publisher).

intersection with the abscissa and is around 180, 220, 230 and 260 K for KCl:Mg^{2+}, KCl:Ca^{2+}, KCl:Sr^{2+}, and KCl:Ba^{2+}, respectively.

The tetragonal distortion resulting from the introduction of the divalent cations into alkali halides is generally formed in the lattice. A dislocation moves on a single slip plane and interacts strongly only with those defects. Then, the relation between the effective stress and the temperature can be approximated as the linear relationship of $\tau^{*1/2}$ vs. $T^{1/2}$ (i.e., the Fleischer's model [1]). τ^* is the effective stress due to the divalent cations. The critical temperature can be also determined from $\tau_{p1}{}^{1/2}$ vs. $T^{1/2}$ for each specimen. The values of T_c are given in **Table 2**. When the divalent ionic size becomes closer to it of K^+ from the small divalent cationic size side, T_c tends to increase. T_c is not influenced by the concentration of additives (i.e., Mg^{2+}, Ca^{2+}, Sr^{2+}or Ba^{2+} here) [12, 15, 16] and is expressed by

Single crystal	Tc (K)	Ionic radius (Å)
KCl:Mg^{2+}	191	Mg^{2+} 0.72
KCl:Ca^{2+}	221	Ca^{2+} 1.00
KCl:Sr^{2+}	227	Sr^{2+} 1.13
KCl:Ba^{2+}	277	Ba^{2+} 1.36
		K^+ 1.38

Table 2.
T_c and ionic radius values for various crystals.

$$T_c = (\Delta G_0/k) \ln \left[\dot{\varepsilon} / \left\{ \rho_{\mathrm{m}} b^2 \nu_D (L_0/L)^2 \right\} \right], \tag{3}$$

where ΔG_0 is the Gibbs free energy for the breakaway of a dislocation from an impurity, ρ_{m} is the density of mobile dislocations, the ν_D is the Debye frequency, and L_0 is the average spacing of divalent cations on a slip plane. At the temperature of T_{c}, thermal fluctuations can provide the entire energy for breaking through the impurity. The additive ions act no longer as obstacles to dislocation motion.

4. Conclusions

The concentration of isolated I-V dipoles affects the τ_{y} values. The values of τ_{y} and ρ remarkably increase with the quenching temperature above 673 K. As for tan δ, it does not vary by quenching from the temperature below 573 K or above 673 K. Within 573 to 673 K, the difference in ρ is slight and the values of tan δ and τ_{y} obviously become larger with a higher quenching temperature. Based on these results, $KCl{:}Sr^{2+}$ single crystals are determined to be quenched from 673 K to room temperature immediately before deformation tests such as compression.

The following two points were mainly mentioned from the experimental results and the discussion based on the data τ_{p1} of the first bending point on the plots of $\Delta\tau$ vs. $(\Delta\tau'/\Delta\ln\dot{\varepsilon})$.

1. The plots of $\Delta\tau$ vs. $(\Delta\tau'/\Delta\ln\dot{\varepsilon})$ have a stair like shape (two bending points and two plateau places) for the KCl doped with the divalent cations. The $\Delta\tau$ values at the first and second bending points, τ_{p1} and τ_{p2}, become obviously larger at lower temperature as well as τ_{y} for the crystals within the temperature.

2. The values of T_{c} were derived from the relation between $\tau_{\mathrm{p1}}^{1/2}$ and $T^{1/2}$ with respect to the Fleischer's model for KCl single crystals doped with Mg^{2+}, Ca^{2+}, Sr^{2+} or Ba^{2+} as divalent impurities. T_{c} tends to increase when the additive cationic size is increasingly close to the K^+ ionic size from the smaller side than K^+ size in the matrix crystal.

Conflict of interest

The author declares no conflict of interest.

Author details

Yohichi Kohzuki
Department of Mechanical Engineering, Saitama Institute of Technology, Fukaya, Japan

*Address all correspondence to: kohzuki@sit.ac.jp

References

[1] Fleischer RL. Rapid solution hardening, dislocation mobility, and the flow stress of crystals. Journal of Applied Physics. 1962;**33**:3504-3508. DOI: 10.1063/1.1702437

[2] Fleischer RL. Solution hardening by tetragonal distortions: Application to irradiation hardening in F.C.C. crystals. Acta Metallurgica. 1962;**10**:835-842. DOI: 10.1016/0001-6160(62)90098-6

[3] Fleischer RL, Hibbard WR. The Relation between Structure and Mechanical Properties of Metal. London: Her Majesty's Stationary Office; 1963. p. 261

[4] Johnston WG, Nadeau JS, Fleischer RL. The hardening of alkali-halide crystals by point defects. Journal of the Physical Society of Japan. 1963; **(Suppl. I)18**:7-15

[5] Sprackling MT. The plastic deformation of simple ionic crystals. In: Alper AM, Margrave JL, Nowick AS, editors. Materials Science and Technology. London New York San Francisco: Academic Press; 1976. p. 93

[6] Lidiard AB. Handbuch der Physik. Vol. 20. Berlin: Springer; 1957. p. 246

[7] Katoh S. Influence of Impurity Concentration on the Blaha Effect (Thesis). Kanazawa: Kanazawa Univ; 1987. pp. 16-21 (in Japanese)

[8] Dryden JS, Morimoto S, Cook JS. The hardness of alkali halide crystals containing divalent ion impurities. Philosophical Magazine. 1965;**12**: 379-391. DOI: 10.1080/14786436508218880

[9] Orozco ME, Mendoza AA, Soullard J, Rubio OJ. Changes in yield stress of NaCl:Pb^{2+} crystals during dissolution and precipitation of solid solutions. Japanese Journal of Applied Physics. 1982;**21**:249-254. DOI: 10.1143/JJAP.21.249

[10] Zaldo C, Solé JG, Agulló-López F. Mechanical strengthening and impurity precipitation behaviour for divalent cation-doped alkali halides. Journal of Materials Science. 1982;**17**:1465-1473. DOI: 10.1007/BF00752261

[11] Reddy BK. Annealing and ageing studies in quenched KBr:Ba^{2+} single crystals. Physica Status Solidi A: Applications and Materials Science. 1987;**99**:K7-K10. DOI: 10.1002/pssa.2210990140

[12] Kohzuki Y, Ohgaku T, Takeuchi N. Interaction between a dislocation and impurities in KCl single crystals. Journal of Materials Science. 1993;**28**:3612-3616. DOI: 10.1007/BF01159844

[13] Kohzuki Y, Ohgaku T, Takeuchi N. Interaction between a dislocation and various divalent impurities in KCl single crystals. Journal of Materials Science. 1995;**30**:101-104. DOI: 10.1007/BF00352137

[14] Kohzuki Y. Study on dislocation-dopant ions interaction in ionic crystals by the strain-rate cycling test during the Blaha effect. Crystals. 2018;**8**:31-54. DOI: 10.3390/cryst8010031

[15] Hikata A, Johnson R. A, Elbaum C. Interaction of dislocations with electrons and with phonons. Physical Review 1970;B2:4856–4863. DOI: 10.1103/PhysRevB.2.4856

[16] Ohgaku T, Takeuchi N. Interaction between a dislocation and monovalent impurities in KCl single crystals. Physica Status Solidi A. 1992;**134**:397-404. DOI: 10.1002/pssa.2211340210

Chapter 6

Elasticity of Auxetic Materials

Jeremiah Rushchitsky

Abstract

The auxeticity of elastic materials is described and explained by the use of the linear and nonlinear models of elastic deformations for a wide range of strain values up to moderate levels. This chapter consists of three parts – general information on auxetic materials, description of auxetics by the model of the linear theory of elasticity, description of auxetics by the models of the nonlinear theory of elasticity. The analytical expressions are offered that corresponds to three kinds of universal deformations (simple shear, uniaxial tension, omniaxial tension) within the frame-work of three well-known in the nonlinear theory of elasticity models – two-constant Neo-Hookean model, three-constant Mooney-Rivlin model, five-constant Murna-ghan model. A most interesting novelty consists in that the sample from elastic material is deformed as the conventional material for small values of strains whereas as the auxetic with increasing to moderate values of strains.

Keywords: auxetic material, elastic deformation of auxetics, three main effects of auxeticity, linear elastic model (Hookean model), nonlinear elastic models (two-constant Neo-Hookean model, three-constant Mooney-Rivlin model, five-constant Murnaghan model, new mechanical effects

1. Introduction

To begin with, let us recall the definition of elasticity of deformation of the material. So, the property of elasticity consists in that the body practically simulta-neously takes the initial configuration after removing the deformation causes. In other words, if deformations are elastic, then they simultaneously vanish after removing the action of forces, caused the deformations.

This property, as also other properties, though, is displayed seldom in the pure form, that is, it is accompanied in real solid materials by several other properties. But in most cases, elasticity is the main and pre-vailing property.

It is worthy to note at beginning of this chapter that the analysis of auxetic materials as the deforming elastically materials is dominating over other types of deformation (thermoelastic, viscoelastic, elastoplastic, magnetoelastic, etc). Therefore, the theme "Elasticity of Auxetic Materials" is related to the main part of studies of auxetics.

At present, the auxetic materials are thought of as some subclass of nontraditional (nonconventional) ma-terials which are known as metamaterials. The metamaterials include the mechanical metamaterials, which in turn include the auxetic materials. At present, a sufficiently big group of scientists work in the area of auxetic materials. It includes mainly specialists from material science, to the lesser extent from statistical physics, and even to the lesser extent from experimen-tal mechanics. The state-of-the-art in science on auxe tic materials is shown in the monographs [1–3] and the review articles [4–23].

The auxetic materials were discovered and identified as a novel class of materials about forty years ago. Usually, two publications of Gibson L.J., Ashby M.F. et al [19, 20] are shown as the pioneer ones.

The term "auxetic material" was introduced by Evans in 1991 [21] for a new range of materials, which he defined them "the materials with negative Poisson ratio (NPR)". This needs some scientific comments relative to the term and definition.

Comment 1 (to the term "auxetic material"). At present, Wikipedia and other sources propose for such materials the name "auxetics". Both names come from the Greek word *αυξητικοζ* (that which tends to increase). But this does not explain why just "auxetic". The next comment on definition gives some clearness.

Comment 2 (to the definition of auxetic material). Auxetic materials are deformed elastically exhibiting the unconventional property of increasing the cross-section (growing swollen) of cylindrical or prismatic samp-le under uniaxial *tension*, whereas in the conventional materials this cross-section decreases (grows thin). Just this is reflected in the name "auxetic" and shown in **Figure 1** [4].

The point is that the property of the decrease is described in the linear theory of elasticity by the use of the Poisson ratio as the elastic constant. A change of the decrease of cross-section on the increase of one means a change of positive values of the Poisson ratio on the negative ones.

The presented short information on auxetics shows that their definition is based on the secondary fact – the negativity of the Poisson ratio, which corresponds to the model of the linearly elastic body. The primary fact consists in observation in the standard for mechanics of materials (which does not depend on the model of deformation) experiment of longitudinal tension of a prism when the transverse deformation of the prism is positive (a material as if swells) in contrast to the classical materials, where it is negative.

The adherence of researchers of auxetic materials to the foams can be seen in the often used (described verbally or by the picture) demonstration of auxeticity of the foam as increasing the volume of sample from the foam under tension. It is shown in **Figure 2** [[10] (left), [20] (right)].

These pictures are really very demonstrative because they show two basic features.

Feature 1. The sample length is possibly not sufficient to create the classical conditions of the test on the universal deformation of uniaxial tension-compression.

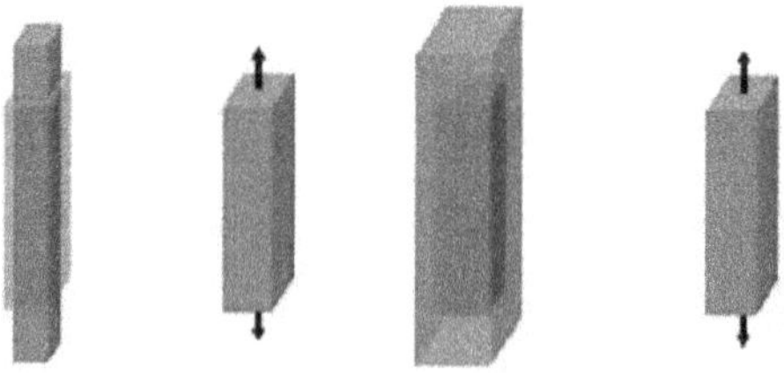

Figure 1.
Test on uniaxial tension for conventional and non-conventional materials.

Figure 2.
Usually used test-demonstration of auxeticity.

Feature 2. The longitudinal and transverse strains are *seemingly* not sufficiently small in this test to be described by the linear theory of elasticity.

Starting with the first works on auxetics, the discussed real materials were the different kinds of foams. It is considered that the first observed auxetic materials were the foams which are characterized by the small value of density and the porous internal structure (see Lakes [24] and Wojciechowski [25]). In the next studies, the new auxetics were revealed, the density of which was also small and which have a porous structure. But it was shown later that small density is not the defining property of auxetics, because the significant part of foams has not the property of auxeticity. The defining characteristics of auxetics are new three mechanical phenomena which will be described below.

The common concept was adopted almost at the initial part of studies that the auxeticity of materials is caused by the internal structure of these materials. This corresponds to the general concept of mechanics (which is clearly shown in mechanics of composite materials) that the specificities of deformation of materials can be explained by the existence of some specific internal structure. Only the answer should be found which concrete specificity is characteristic for the auxetic materials. Therefore, an essential part of studies of auxetics consists in the finding of diverse variants of internal structure that are further studied by methods of molecular physics and computational simulations. The most popular is a so-called hexagonal system (it is shown in **Figure 3** [12]; left – before stretching, right – after stretching). Just this structure shows the swelling of the sample and is given by different authors to illustrate the auxeticity.

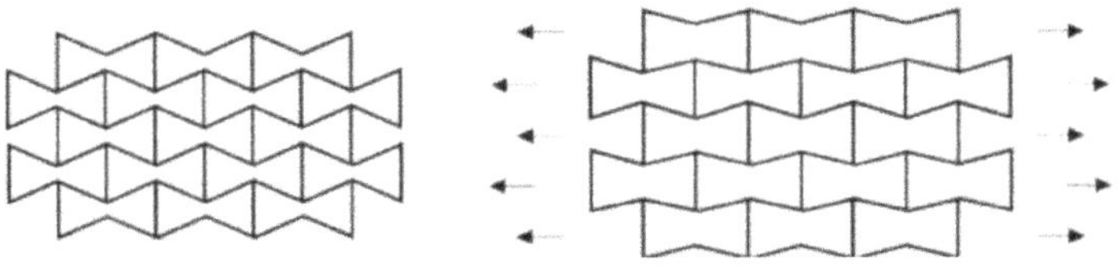

Figure 3.
The most known interpretation of the internal structure of auxetic material.

It should be noted that the mechanics of materials works with the continuum models. This means that any discrete models of the internal structure must be transformed into the continuum one (here the different ways of averaging are usually applied). In mechanics, the internal structure of materials can appear on two different stages of modeling the materials. First, on the stage of changing the discrete structure of a material by the continuous one (that is when the notion of the continuum is introduced according to the principle of continualization). Second, on the stage of modeling the piece-wise inhomogeneous continuum by the homo-geneous continuum (that is when the principle of homogenization is applied). The first stage is usually asso-ciated with methods of molecular physics, whereas the second stage is a standard one in mechanics of composite materials. This is peculiar to all the materials that are studied in mechanics and refers also to the theory of elasticity within the framework of which the elastic deformation of aux-etic materials is studied.

For the presence in the material property of auxeticity, its internal structure has to change under defor mation by the special way exhibiting the unusual (nontraditional) mechanical effects. Note that mechanics of materials studied traditionally first the elastic deformation and this concern both traditional (non-auxe-tic), and nontraditional (auxetic) materials.

As far as the number of known nonauxetic materials exceeds the number of auxetic ones on many or-ders, then the term "unusual effect" is looking

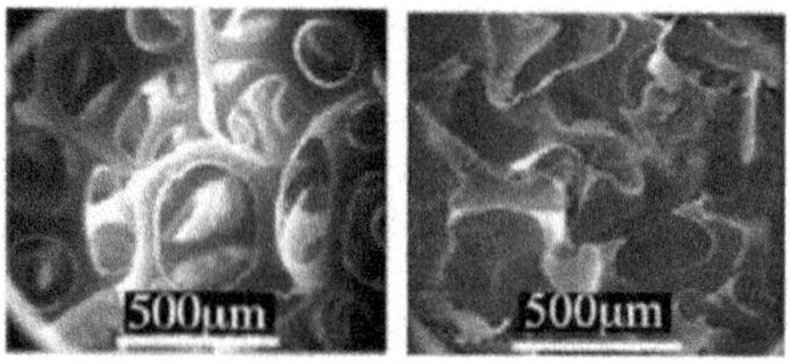

Figure 4.
Sample from the polyurethane foam (left – traditional structure, right – auxetic structure).

appropriate. In contrast to the traditional effects that count tens, the effects of auxeticity are observed as now in the identical mechanical problems in three types of such problems that are realized experimentally and described theoretically. An identity consists in that the samples from material must be compared when the internal structure of a material in cases "auxetic non-auxetic" is differing by the only geometrical shape of pores. This case is shown in **Figure 4** for the sample from the polyurethane foam (left – traditional structure, right – auxetic structure) [2].

Now, some facts from this theory should be shown concerning the phenomenon of auxeticity. But first three specific appearances of auxeticity must be described and commented on.

2. Three specific properties of auxetic materials

Only one of these specificities is well known – the swelling under the tension of the standard sample (standard mechanical test). This test is described above and shown in **Figure 1**.

But the fact is known that the auxeticity is generated by the special kind of internal structure of material and appears in three basic mechanical tests on deformation of material

1. *Swelling under tension.*

2. *Hardening under indentation (impact).*

3. *Synclastic* and *anticlastic deformation of thin flexible plate.*

Test 2 on indentation (statical Hertz problem, problem on hardness by Rockwell-Brinell-Wikkers) and impact (dynamical Hertz problem) shows the effect of hardness of auxetics in the contact zone. Within the framework of the theory of elasticity, this problem is solving numerically with the given exactness. A scheme of test that exhibits the essential difference in the degree of indentation of the spherical indentor into the traditional (left) and auxetic (right) materials is shown in **Figure 5** [12].

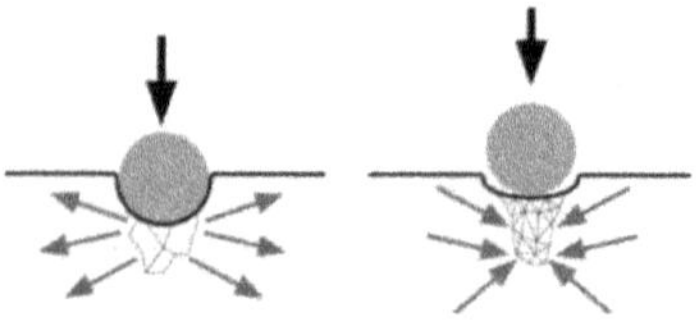

Figure 5.
Test for hardness material (left – traditional structure, right – auxetic structure).

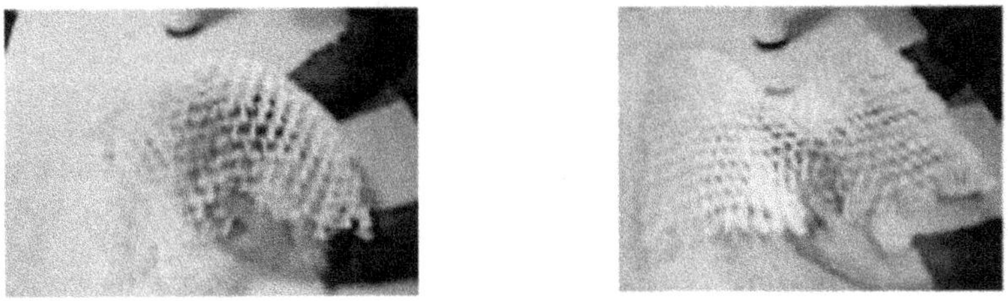

Figure 6.
Test on synclastic (left) and anticlastic (right) deformation.

Test 3 on synclastic and anticlastic deformation of flexible elastic plate is stated within the assumption that the plate is quadratic in plan and is loaded by the balanced system of three forces – one force is applied at the center of a plate and directed upward, whereas two other identical forces are applied at the centers of two opposite ends of the plate and directed downward. Within the framework of the theory of flexible elastic plates, this problem is solving numerically with the given exactness. The simple experiment that exhibits the essential difference in deformation of the plate from the traditional and auxetic materials is shown in **Figure 6** [4] (left traditional material, right – auxetic material).

Note that these basic phenomena of deformation of auxetics can be described only in the terms of the theory of elasticity.

3. Some facts from the linear theory of elasticity necessary for describing the auxetic materials

Because the elastic deformation is described in mechanics only by the theory of elasticity, then some facts from this theory should be recalled before the discussion of the specificities of elastic deformation of auxetics. At that, the division of the theory of elasticity on the linear and nonlinear theories should be taken into account. It is important to remember that the linear theory is based on the one (Hookean) model, whereas the nonlinear uses many different models.

3.1 Universal deformations

Universal deformations (uniform deformations, universal states) occupy a special place in the theory of elasticity just owing to their universality [26]. This universality consists in that the theoretically and experi-mentally determining elastic constants of material in samples, in which the universal deformation is created purposely, are valid also for all other deformed states both samples and any different products made of this material. It is considered therefore that the particular importance of universal deformation (their fundamen tality) consists in the possibility to use them in the determination of properties of materials from tests [26–31]. To realize the universal deformation, two conditions have to be fulfilled: 1. Uniformity of deformation must not depend on the choice of material. 2. Deformation of material has to occur by using only the surface loads.

In the theory of infinitesimal deformations, the next kinds of universal deformations are studied more of-ten and in detail: simple shear, simple (uniaxial) tension-compression, uniform volume (omniaxial) tension-compression. In the linear theory of elasticity, the experiment with a sample, in which the simple shear is realized, allows determining the elastic shear modulus μ. The experiment with a sample, in which the uniaxial tension is realized, allows determining Young elastic modulus E and Poisson ratio ν. The experiment with a sample, in which the uniform compression is realized, allows determining the elastic bulk modulus k.

While being passed from the linear model, which is valid for only the very small deformations to the mo-dels of non-small (moderate or large) ones, that is, from the linear mechanics of materials to nonlinear me-chanics of materials, the universal states permit to describe theoretically and experimentally many nonlinear phe-nomena. The history of mechanics testifies to the experimental observation in the XIX century of the non linear effects that arose under the simple shear and were named later by the names of Poynting and Kelvin [27–31]. After about a hundred years in the XX century, these effects were described theoretically within the framework of the nonlinear Mooney-Rivlin model [31–35].

The mechanics of composite materials is one more area of application of universal deformations. The mo-del of averaged (effective, reduced) moduli is in this case the simplest and most used model. In the theory of effective moduli, the composite materials of the complex internal structure with internal links are treated usually as homogeneous elastic media. A possibility to create in such media the states with universal deformations was used in the evaluation of effective moduli by different authors and different methods. It was found that it is sufficient for isotropic (granular) composites to study the energy stored in the elementary volumes of composites under only two kinds of universal deformations: simple shear and omniaxial compression. In the case of transversely isotropic (fibrous or layered) composites, the different directions need analysis of universal deformations for each direction separately.

3.2 Classical procedures of estimating the values of elastic moduli in the linear theory of elasticity

Perhaps, the eldest and exhausting procedures are shown in the classical Love's book [36]. Let us save the Love's notations and write according to [36] the internal energy of deformation of the linearly elastic isotropic body W in the form

$$W = \lambda(\varepsilon_{xx} + \varepsilon_{yy} + \varepsilon_{zz})^2 + 2\mu\left(\varepsilon_{xx}^2 + \varepsilon_{yy}^2 + \varepsilon_{zz}^2\right) + \mu\left(\varepsilon_{xy}^2 + \varepsilon_{xz}^2 + \varepsilon_{yz}^2\right), \tag{1}$$

where λ, μ are the Lame moduli, $\varepsilon_{xx}, \dots, \varepsilon_{yz}$ are the components of the strain tensor. The Hooke law has the form

$$\begin{aligned} X_x &= \lambda\Delta + 2\mu\varepsilon_{xx}, Y_y = \lambda\Delta + 2\mu\varepsilon_{yy}, Z_z = \lambda\Delta + 2\mu\varepsilon_{zz}, \\ X_y &= 2\mu\varepsilon_{xy}, Z_x = 2\mu\varepsilon_{zx}, Y_z = 2\mu\varepsilon_{yz}. \end{aligned} \tag{2}$$

Here $\Delta = \varepsilon_{xx} + \varepsilon_{yy} + \varepsilon_{zz}$ is the dilatation.

The classical procedure of introducing the Young modulus and Poisson ratio is as follows: the cylinder or prism of any shape is considered, then the axis of the cylinder is chosen in direction Ox and the prism is stretched at the ends by a uniform tension T. Because the lateral surface of the prism is assumed to be free of stresses, then the stress state of a prism is uniform and is characterized by only one stress $X_x = T$. In this case, the Hooke law becomes simpler

$$T = \lambda\Delta + 2\mu\varepsilon_{xx}, 0 = \lambda\Delta + 2\mu\varepsilon_{yy}, 0 = \lambda\Delta + 2\mu\varepsilon_{zz}. \tag{3}$$

An expression for dilatation follows from equalities (3) $T = (3\lambda + 2\mu)\Delta \rightarrow \Delta = T/(3\lambda + 2\mu)$.

The substitution of the last expression for dilatation into the first equality (2) gives relations

$$T = \frac{\lambda}{3\lambda + 2\mu} T + 2\mu\varepsilon_{xx} \rightarrow T = \frac{\mu(3\lambda + 2\mu)}{\lambda + \mu}\varepsilon_{xx}. \tag{4}$$

The expression (4) represents the elementary law $T = E\varepsilon_{xx}$ of link between tension and deformation of the prism, in which the Young modulus E is used

$$E = \frac{\mu(3\lambda + 2\mu)}{\lambda + \mu}. \tag{5}$$

The substitution of expression for dilatation into the second and third equalities (2) gives relations

$$-\varepsilon_{yy} = -\varepsilon_{zz} = \frac{\lambda}{2(\lambda + \mu)}\varepsilon_{xx}, \tag{6}$$

which express the classical Poisson law on the transverse compression under the longitudinal extension and permit to introduce of the Poisson ratio

$$\sigma = \frac{-\varepsilon_{yy}}{\varepsilon_{xx}} = \frac{-\varepsilon_{zz}}{\varepsilon_{xx}} = \frac{\lambda}{2(\lambda + \mu)}. \tag{7}$$

Let us repeat now the procedure associated with introducing the universal (uniform) deformation – the uniform compression. Thus, the body of arbitrary shape is considered, to all points of which the constant pressure $-p$ is applied. In this body, the uniform stress state arises which is characterized by stresses $X_x = Y_y = Z_z = -p$, $X_y = Y_z = Z_x = 0$. The Hooke law becomes simpler

$$-p = \lambda\Delta + 2\mu\varepsilon_{xx}, \ -p = \lambda\Delta + 2\mu\varepsilon_{yy}, \ -p = \lambda\Delta + 2\mu\varepsilon_{zz}, \varepsilon_{xy} = \varepsilon_{yz} = \varepsilon_{zx} = 0. \tag{8}$$

The relations (8) can be transformed to $-3p = (3\lambda + 2\mu)(\varepsilon_{xx} + \varepsilon_{yy} + \varepsilon_{zz}) \rightarrow -p = [\lambda + (2/3)\mu]\Delta$.

In this way, the modulus of compression k is defined

$$k = \lambda + (2/3)\mu. \tag{9}$$

The classical Love's reasoning, which is repeated in most books on the linear theory of elasticity, is based on the representation of moduli λ, μ, k through moduli E, σ

$$\lambda = \frac{E\sigma}{(1+\sigma)(1-2\sigma)}, \ \mu = \frac{E}{2(1+\sigma)}, \ k = \frac{E}{3(1-2\sigma)}. \tag{10}$$

The formulas (10) are commented in ([36], p. 104) as follows: "If σ were $> 1/2$, k would be negative, or the material expands under pressure. If σ were < -1, μ would be negative, and the function W would not be a positive quadratic function. We may show that this would also be the case if k were negative. Negative values σ are not excluded by the condition of stability, but such values have not been found for any isotro-pic material."

Because the comments of negativity of Poisson ratio is found in the books on the theory of elasticity very seldom, therefore a few sentences from Lurie's book ([29],

p. 117) are worthy to be cited: "A tension of the rod with negative ν (but the more than -1) would be accompanied by increasing of transverse sizes. Ener-getically, the existence of such elastic materials is not excluded." "In hypothetic material with $\nu < -1$, the hy-drostatic compression of the cube would be accompanied by increasing its volume."

Note that the Poisson ratio is denoted in the theory of elasticity by σ "sigma" and ν "nu". Love uses σ, whereas Lurie uses ν.

It should be also noted that not all authors of books on the linear isotropic theory of elasticity discuss the restrictions on changing the Poisson ratio (for example, Germain, Nowacki, Hahn do not made this in their well-known books [37–39]). The constitutive relations and classical restrictions on elastic constants are discussed in the most comprehensive and modern treatment of the theory of elasticity [28] (Subsection 3.3 "Constitutive relations").

But in some books, the discussion is presented and all authors start with one and the same postulate: in the procedure of restrictions in changing the Poisson ratio, the primary requirement is a positiveness of internal energy W (1). The representation of energy can be different for different elastic moduli. For example, Leibensohn [40], Love [36], Lurie [29] choose the pair λ, μ and use the representation (1). Landau and Lifshits [41] use the pair k, μ. In all the cases, W has a form of a quadratic function with coefficients composed of elastic moduli.

Thus, in most cases, the expression (1) is analyzed. It is assumed that the sufficient and being in line with experimental observations condition is the condition of positiveness of Lame moduli

$$\lambda > 0, \mu > 0. \tag{11}$$

Further, the formulas (10) are considered, in which without controversy the Young modulus is assumed positive $E > 0$. Then positiveness of expressions $1 + \sigma > 0, \ 1 - 2\sigma > 0$ provides validity of formula (11), from which the well-known restriction on the Poisson ratio follows

$$-1 < \sigma < 1/2. \tag{12}$$

Let us recall that all the elastic moduli in the classical linear isotropic theory of elasticity are always posi-tive. The obvious contradiction between the assumption of negativity of the Poisson ratio and the primary statement on the positivity of Lame moduli (11) in condition when the Poisson ratio is defined by formula (7) is commented in the classical theory of elasticity anybody. To all appearances, this situation is occurred owing to the incredibility of negative values if only one of the elastic moduli λ, μ, E, k.

Note finally that two experimental approaches to determine the value of Poisson ratio for concrete material are used at present time ([27], subsections 2.18, 3.27, 3.28). The first approach is the older one. It is based on the experimental determination of Young, shear, and compression moduli and subsequent calculation of Poisson ratio by formulas (10) $\sigma = (E/2\mu) - 1, \ \sigma = (1/2)[(E/3k) - 1]$. Here, the problem of the exactness of calculation arises. Let us cite Bell's book ([27], subsection 3.28): "Remind of the Grüneisen's conclusion that the errors of $\pm 1\%$ in values E and μ result in the error of 10% in the value of Poisson ratio." Therefore, the second approach seems to be more preferable. It is associated with Kirchhoff's experiments (1859), in which the Poisson ratio is determined from the direct experiment on simultaneous bending and torsion.

Let us recall that the primary phenomenon in the determination of the Poisson ratio is the contraction of a sample (transverse deformation of a sample) under its elongation (its longitudinal deformation).

3.3 Refinement of procedures of estimating the values of elastic moduli

Let us save the initial postulate that the primary requirement is the positivity of internal energy W (1) and reject the sufficient (and not necessary) condition of positivity of W when the positivity of Lame moduli λ, μ is assumed and suppose the general condition of positivity of W.

Because the Lame modulus μ has a physical sense of the shear modulus and until now the facts of observation its negativity (a shear in the direction opposite to the direction of shear force) are not reported, then we can agree to its positivity. This condition of positivity can be also substantiated theoretically based on ana-lysis of universal deformation of simple shear. To describe the simple shear, the coordinate plane (for example, xOy) should be chosen and only one non-zero component $u_{x,y}$ of the displacement gradient should be given. This can be commented geometrically as deformation of the elementary rectangle $ABCD$ with sides dx, dy parallel to the coordinate axes into the parallelogram $AB'C'D$, which results from the longitudinal shift of the rectangle side BC. Then the shear angle $\measuredangle BAB' = \gamma$ is linked with $u_{x,y}$ in a next way $u_{x,y} = \tan\gamma = \tau$ and $\varepsilon_{xy} = (1/2)\tau$. The Hooke law becomes the simplest form $\sigma_{xy} = 2\mu\varepsilon_{xy}$ and the corresponding representation of internal energy is as follows $W = (1/2)\mu\tau^2$. Then the positivity of shear modulus (14) follows from the positivity of energy W.

Now, the next refinement can be formulated.

Refinement 1. The Lame modulus λ can be negative if the Poisson ratio $\sigma = \lambda/[2(\lambda+\mu)]$ can be assumed possible negative.

Refinement 2. If the Poisson ratio σ is assumed to be possible negative and the shear modulus μ is positive, then according to definition (7) the negative Lame modulus λ can not exceed by its absolute value the shear modulus

$$|\lambda| < \mu. \tag{13}$$

Let us return to the primary definition of the Poisson ratio (7), which is found from the solution of the problem of unilateral tension. In this case, the internal energy has the form

$$\begin{aligned} W &= \lambda(\varepsilon_{11}+\varepsilon_{22}+\varepsilon_{33})^2 + 2\mu\left(\varepsilon_{11}^2+\varepsilon_{22}^2+\varepsilon_{33}^2\right) = \lambda(\varepsilon_{11}+2\sigma\varepsilon_{11})^2 + 2\mu\left(\varepsilon_{11}^2 + (\sigma\varepsilon_{11})^2 + (\sigma\varepsilon_{11})^2\right) \\ &= \left[\lambda(1+2\sigma)^2 + 2\mu\left(1+2\sigma^2\right)\right]{\varepsilon_{11}}^2 > 0. \end{aligned} \tag{14}$$

Then $\lambda + \left[2(1+2\sigma^2)/(1+2\sigma)^2\right]\mu > 0.$ permits to the formulation of some new refinements.

Refinement 3. If the Poisson ratio σ is assumed to be possible negative and the shear modulus μ is positive, then the condition of positivity of internal energy admits arbitrary negative values of the Poisson ratio.

(because the coefficient ahead of μ is always positive). The case $1+2\sigma = 0 \to \sigma = -0,5$ is the peculiar one – the value of modulus λ is practically not restricted at its neighborhood.

Refinement 4. The Lame modulus λ is already restricted from below according to (14), but also the additional condition (16) exists

$$|\lambda| < \left[2(1+2\sigma^2)/(1+2\sigma)^2\right]\mu. \tag{15}$$

The condition (15) is less strong: the coefficient ahead of μ exceeds 1 for all negative σ (in condition (13), the coefficient ahead of μ is equal to 1). Therefore, the condition (13) remains.

Let us turn to formula (9), which expresses the compression modulus k through the Lame moduli λ, μ. It follows from (9) that the modulus k will be negative if only the negative Lame modulus λ exceeds $(2/3)\mu$ by absolute value

$$k = \lambda + (2/3)\mu < 0 \rightarrow |\lambda| > (2/3)\mu = 0,667\mu. \quad (16)$$

Comparison with restrictions (13) and (15) on the absolute values of negative Lame modulus λ in the case of negative values of Poisson ratio σ shows that (16) does not conflict with (13) and (15).

Refinement 5. If the Poisson ratio σ is assumed to be possible negative and the shear modulus μ is positive, then the compression modulus k can be negative.

The situation with refinements becomes clearer if the moduli λ, E and k are written through μ and σ

$$\lambda = \frac{2\sigma}{1-2\sigma}\mu, E = 2(1+\sigma)\mu, k = \frac{2}{3}\frac{1+\sigma}{1-2\sigma}\mu. \quad (17)$$

A few statements can be formulated at the end of this subchapter.

Statement 1. The classical restrictions of positivity of the elastic moduli in the isotropic theory of elasticity should be refined for auxetic materials: most elastic moduli can be negative.

Statement 2. Seemingly, the auxetics should be defined by the primary physical phenomenon of positivity of transverse deformation of a prism, which is observed in the standard in mechanics of materials experiment of longitudinal tension of a prism. In this case, the auxetics will be associated not only with the isotropic elastic materials.

Statement 3. In the case of auxetic materials, the Lame modulus λ is always negative and the Young E and compression k moduli are negative when the negative Poisson ratio is less than -1: $\sigma < -1$.

Statement 4. When the problems of the linear isotropic theory of elasticity being studied for auxetic materials, then at least two elastic moduli for these materials should be determined from the direct experiments (unilateral tension, omnilateral compression, simple shear, torsion).

4. Specificities of describing the auxetic materials by the nonlinear theory of elasticity

4.1 Essentials of nonlinear theory of elasticity

While being studied the auxetics from the position of the nonlinear theory of elasticity, some essential differences between the linear and nonlinear descriptions should be taken into account. Therefore, the basic notions of the nonlinear approach seem to be worthy to show here very shortly [31, 34, 35, 42, 43].

A body is termed some area V of 3D space R^3, in each point of which the density of mass ρ is given (the area occupied by the material continuum). In this way, a real body, the shape of which coincides with V, is changed on a fictitious body. This fictitious body is the basic notion of mechanics. The Lagrangian $\{x_k\}$ or Eulerian $\{X_k\}$ coordinate systems can be given in R^3. In the theory of deformation of a body as a change of its initial shape, the notions are utilized that are associated with a geometry of body (kinematic notions) and with the forces acting on the body from outside and inside (kinetic notions). The notions of the configuration χ, the vector of displacement $\vec{u} = \{u_k\}$, the principal extensions λ_k, the strain tensor ε_{ik} are referred to as the notions of kinematics. The external and internal forces, as well as the tensors of internal stresses, refer to the notions of kinetics,

The configuration of the body at a moment t is called the actual one, whereas the configuration of the body at arbitrarily chosen initial moment t^o is called the reference one. The coordinates of the body point before deformation are denoted by x_k. It is assumed that after deformation this point is displaced on the va-lue $u_k(x_1, x_2, x_3, t)$. Then the vector with components u_k is called the displacement vector and the coordinates of the point after deformation are presented in the form $\xi_k = x_k + u_k(x_1, x_2, x_3, t)$. The frequently used Cauchy-Green strain tensor is given by the known displacement vector $\vec{u}(x_k, t)$ in the Lagrangian coordinates $\{x_k\}$ and the reference configuration

$$\varepsilon_{nm}(x_k, t) = (1/2)(u_{n,m} + u_{m,n} + u_{n,i}u_{m,i}). \tag{18}$$

As a result, the deformation of the body is given by nine components of displacement gradients $u_{i,k}$. Such a description of deformation is used in most models of the nonlinear theory of elasticity. But the process of deformation can be described also by other parameters of the geometry change of the body. It seems meaning ful to use often the first three algebraic invariants of tensor (18) $A_1 = \varepsilon_{mn}\delta_{mn}$, $A_2 = (1/2)\left[(\varepsilon_{mn}\delta_{mn})^2 - \varepsilon_{ik}\varepsilon_{ik}\right]$, $A_3 = \det \varepsilon_{mn}$, which can be rewritten through the principal values of tensor (18) ε_k by the formulas $A_1 = \varepsilon_1 + \varepsilon_2 + \varepsilon_3$, $A_2 = \varepsilon_1\varepsilon_2 + \varepsilon_1\varepsilon_3 + \varepsilon_2\varepsilon_3$, $A_3 = \varepsilon_1\varepsilon_2\varepsilon_3$. The often used invariants I_1, I_2, I_3 of tensor ε_{ik} are linked with the algebraic invariants of the same tensor by relations

$$I_1 = 3 + 2\varepsilon_{nn} = 3 + 2A_1,\ \ I_2 = 3 + 4\varepsilon_{nn} + 2(\varepsilon_{nn}\varepsilon_{mm} - \varepsilon_{nm}\varepsilon_{mn}) = 3 + 4A_1 + 2(A_1^2 - A_2),$$
$$I_3 = \det\|\delta_{pq} + 2\varepsilon_{pq}\| = 1 + 2A_1 + 2(A_1^2 - A_2) + (4/3)(2A_3 - 3A_2A_1 + A_1^3).$$

In several models of nonlinear deformation of materials, the elongation coefficients (principal extensions) defined as a change of length of the conditional linear elements (the infinitesimal segments that are directed arbitrarily) are used

$$\lambda_k = \sqrt{1 + 2\varepsilon_k}. \tag{19}$$

A simpler formula $\lambda_k - 1 \approx \varepsilon_k$ is valid for the case of linear theory. Additionally to three parameters (19), three parameters should be introduced that characterize a change of the angles between linear elements and areas of elements of coordinate surfaces.

It seems to be necessary to show the very often used notation of the displacement gradient

$$\mathbf{F} = \begin{bmatrix} 1 + u_{1,1} & u_{1,2} & u_{1,3} \\ u_{2,1} & 1 + u_{2,2} & u_{2,3} \\ u_{3,1} & u_{3,2} & 1 + u_{3,3} \end{bmatrix}$$

and notation of the left Cauchy-Green strain tensor $\mathbf{B} = \mathbf{F}\,\mathbf{F}^{\mathrm{T}}$ associated with it. The most used are two ten-sors of internal stresses: the symmetric Cauchy-Lagrange tensor σ_{ik}, which is measured on the unit of area of the deformed body, and the nonsymmetric Kirchhoff tensor t_{ik}, which is measured on the unit area of the undeformed body.

4.2 Universal deformation of simple shear in the nonlinear approach

The simple shear is described in subsubsection 3.3, where the basic formula $u_{1,2} = \tan\gamma = \tau > 0$ is shown.

In the linear theory, the shear angle is assumed to be small and then $\gamma \approx \tan\gamma = \tau$. The nonlinear app-roach introduces some complications. The Cauchy-Green strain tensor is characterized by only three non-zero components

$$\begin{aligned} \varepsilon_{11} &= (1/2)(u_{1,1} + u_{1,1} + u_{1,k}u_{1,k}) = (1/2)(u_{1,2}u_{1,2} + u_{1,3}u_{1,3}) = \tau^2; \varepsilon_{12} = \varepsilon_{21} \\ &= (1/2)(u_{1,2} + u_{2,1} + u_{1,k}u_{2,k}) = (1/2)\tau, \end{aligned} \tag{20}$$

The principal extensions are written through the shear angle by formulas $\lambda_1 = 1,\ \lambda_2 = \lambda_3 = \tau$.

4.3 Universal deformation of uniaxial tension in the nonlinear approach

This kind of deformation is also described above. It is characterized in the nonlinear approach by only one nonzero component σ_{11} of the stress tensor and two nonzero components $\varepsilon_{11}, \varepsilon_{22} = \varepsilon_{33}$ of the strain tensor (or two principal extensions $\lambda_1, \lambda_2 = \lambda_3$).

4.4 Universal deformation of uniform (omniaxial) compression-tension

A sample has the shape of a cube, to sides of which the uniform surface load (hydrostatic compression) is applied. Then the uniform stress state is formed in the cube. The normal stresses are equal to each other $\sigma_{11} = \sigma_{22} = \sigma_{33}$, and the shear stresses $\sigma_{ik} (i \neq k)$ are absent. This type of deformation is defined as follows

$$u_{1,1} = u_{2,2} = u_{3,3} = \varepsilon > 0, u_{1,1} + u_{2,2} + u_{3,3} = 3\varepsilon = e, u_{k,m} = (\partial u_k / \partial x_m) = 0 \ \ (k \neq m). \tag{21}$$

The Cauchy-Green strain tensor is simplified $\varepsilon_{11} = \varepsilon_{22} = \varepsilon_{33} = \varepsilon + (1/2)\varepsilon^2,\ \ \varepsilon_{ik} = 0 \ \ (i \neq k)$ and the algebraic invariants are written in the form

$$I_1 = \varepsilon_{11} + \varepsilon_{22} + \varepsilon_{33} = e, I_2 = (\varepsilon_{11})^2 + (\varepsilon_{22})^2 + (\varepsilon_{33})^2, I_3 = (\varepsilon_{11})^3 + (\varepsilon_{22})^3 + (\varepsilon_{33})^3. \tag{22}$$

The principal extensions are equal to each other

$$\lambda_1 = \lambda_2 = \lambda_3. \tag{23}$$

4.5 Three nonlinear models of hyperelastic deformation

These models are related to the models of hyperelastic materials. This class of materials is characterized by the way of introduction of constitutive equations. First, the function of kinematic parameters (elastic potential, internal energy) is defined, from which later the constitutive equations are derived mathematically and sub-stantiated physically. Model 1 is chosen as the simplest one. Model 2 is well-working for the not-small (large or finite) deformations. Model 3 belongs to the most used in the nonlinear mechanics of materials.

4.5.1 Two-constant Neo-Hookean model (model 1)

The elastic potential of this model is defined as follows [31, 34, 35, 42, 43]

$$\begin{gathered} W = C_1(\bar{I}_1 - 3) + D_1(J - 1)^2, \quad \bar{I}_1 = J^{-2/3} I_1, \ J = \det u_{i,k}, \\ W(\lambda_1, \lambda_2, \lambda_3) = C_1\left[(\lambda_1\lambda_2\lambda_3)^{-2/3}(\lambda_1^2 + \lambda_2^2 + \lambda_3^2) - 3\right] + D_1(\lambda_1\lambda_2\lambda_3 - 1)^2. \end{gathered} \tag{24}$$

Here the elastic constants of the model are linked with the classical elastic constants by relation $2C_1 = \mu$; $2D_1 = k$.

The constitutive equations have the form

$$\sigma_{nm} = 2C_1 J^{-5/3}[B_{nm} - (1/3)I_1\delta_{nm}] + 2D_1(J-1)\delta_{nm} \tag{25}$$
$$\sigma_{nn} = 2C_1 J^{-5/3}(\lambda_n - (1/3)I_1) + 2D_1(J-1).$$

It is considered that this model describes well the deformation of rubber under the principal extensions up to 20% from the initial state. Since these extensions are linked with the principal values of the strain ten-sor by relation $\lambda_k = \sqrt{1+2\varepsilon_k}$, then it is assumed $\lambda_k - 1 \approx \varepsilon_{kk}$ approximately with exactness to $\leq 1\%$ in the cases of universal deformations for Neo-Hookean model, what is true in the case of linear theory too. Because the extensions in the linear theory are two orders less, then this observation testifies the fact that the Neo-Hookean model extends essentially the area of allowable values of strains as compared with the Hookean model.

4.5.2 Three-constant Mooney-Rivlin model (model 2)

The elastic potential of the Mooney -Rivlin model is defined as follows [31–35, 42, 43]

$$W = C_{10}(\bar{I}_1 - 3) + C_{01}(\bar{I}_2 - 3) + D_1(J-1)^2, \ \bar{I}_2 = J^{-4/3}I_2, \tag{26}$$
$$W(\lambda_1, \lambda_2, \lambda_3) = C_{10}\left[(\lambda_1\lambda_2\lambda_3)^{-2/3}(\lambda_1^2 + \lambda_2^2 + \lambda_3^2) - 3\right]$$
$$+C_{01}\left[(\lambda_1\lambda_2\lambda_3)^{-4/3}(\lambda_1^2\lambda_2^2 + \lambda_1^2\lambda_2^2 + \lambda_2^2\lambda_3^2) - 3\right] + D_1(\lambda_1\lambda_2\lambda_3 - 1)^2,$$

where the elastic constants are linked with the classical constants by relations $2(C_{10} + C_{01}) = \mu$; $2D_1 = k$.

The stresses are determined by formulas

$$\begin{aligned}\sigma = 2J^{-5/3}(C_{10} + C_{01}J^{-2/3}I_1)B - 2J^{-7/3}C_{01}BB \\ +\left[2D_1(J-1) - (2/3)J^{-5/3}(C_{10}I_1 + 2C_{01}J^{-2/3}I_2)\right]1,\end{aligned} \tag{27}$$

$$\begin{aligned}\sigma_{kk} &= \lambda_k \frac{\partial W}{\partial \lambda_k} \\ &= 2C_{10}(\lambda_1\lambda_2\lambda_3)^{-5/3}\left[\lambda_k^2 - (1/3)(\lambda_1^2 + \lambda_2^2 + \lambda_3^2)\right] \\ &\quad +2C_{01}(\lambda_1\lambda_2\lambda_3)^{-7/3}\left[\lambda_k^2(\lambda_n^2 + \lambda_m^2) - (2/3)\lambda_k(\lambda_1^2\lambda_2^2 + \lambda_1^2\lambda_2^2 + \lambda_2^2\lambda_3^2)\right] + D_1(\lambda_1\lambda_2\lambda_3 - 1).\end{aligned} \tag{28}$$

Here the indexes *knm* form the cyclic permutation from numbers 123.

The Mooney-Rivlin model is the classical one. This can be seen from the next historical information.

Information. An effect of nonlinear dependence of decreasing the shear stresses when the torsion angle (de-formation) to the level of nonsmall values is called "the Poynting effect" owing to his publication of 1909, where this effect was described. At that, Poynting does not mention the results of Coloumb (1784), Wert-heim (1857), Kelvin (1865), Bauschinger (1881), Tomlinson (1883), where this effect was also described in one way or another. But only within the framework of finite elastic deformations, which was developed in 20 century, this effect was satisfactorily explained by Rivlin in 1951. He used the model of nonlinear defor-mation which now is termed “the Mooney-Rivlin model”.

4.5.3 *Five-constant Murnaghan model (model 3)*

The elastic potential in the Murnaghan model has the form [31, 34, 35, 42–45]

$$W(\varepsilon_{ik}) = (1/2)\lambda(\varepsilon_{mm})^2 + \mu(\varepsilon_{ik})^2 + (1/3)A\varepsilon_{ik}\varepsilon_{im}\varepsilon_{km} + B(\varepsilon_{ik})^2\varepsilon_{mm} + (1/3)C(\varepsilon_{mm})^3, \quad (29)$$

$$W(I_1, I_2, I_3) = (1/2)\lambda I_1^2 + \mu I_2 + (1/3)AI_3 + BI_1I_2 + (1/3)I_1^3.$$

The Cauchy-Green strain tensor ε_{ik} and five elastic constants (two Lame elastic constants λ, μ and three Murnaghan elastic constants A, B, C) are used in this potential.

The Murnaghan model can be considered as the classical one in the nonlinear theory of hyperelastic ma-terials. It takes into account all the quadratic and cubic summands from the expansion of the internal energy and describes the deformation of a big class of engineering and other materials. If to unite the data on the constants of Murnaghan model, shown in books [21, 42, 44], then the sufficiently full information can be ob-tained on many tens of materials.

5. Description of deformations of the auxetic materials by the models 1–3

5.1 Universal deformation of simple shear

This kind of deformation of the auxetics needs some preliminary discussion. First, mechanics distinguishes the simple and pure shears. The state of such defor-mations is standard in the test for the determination of the shear modulus. Second, it is a common position in mechanics that this modulus is always positive. This means that new effects relative to auxetic materials will most likely not be found. Third, owing to the written above comments, the one only positive result can be reached: the degree of the description of the classical nonlinear effects the Poynting and Kelvin effects – can be considered for the chosen three nonlinear models.

The following materials are used in the numerical evaluations below (elastic constants are shown): 1. Rubber - $\mu = 20$ MPa, $k = 2.0$ GPa. 2. Foam - $\lambda = 0.58 \cdot 10^9$, $\mu = 0.39 \cdot 10^9$, $k = 0.84 \cdot 10^9$. 3. Foam - $\lambda = 0.58 \cdot 10^9$, $\mu = 0.39 \cdot 10^9$, $A = -1.0 \cdot 10^{10}$, $B = -0.9 \cdot 10^{10}$, $C = -1.1 \cdot 10^{10}$. 4. Polystyrene - $\lambda = 3.7 \cdot 10^9$, $\mu = 1.14 \cdot 10^9$, $A = -1.1 \cdot 10^{10}$, $B = -0.79 \cdot 10^{10}$, $C = -0.98 \cdot 10^{10}$.

5.1.1 *Simple shear in model 1*

In this case $J = (1+\tau)^2$, $I_1 = 1 + 2\tau^2$. Then expressions for displacement gradients **F** and components of tensor **B** are simplified

$$\mathbf{F} = \begin{bmatrix} 1 & \tau & \tau \\ 0 & 1 & 0 \\ 0 & 0 & 1 \end{bmatrix}, \mathrm{B} = \begin{bmatrix} 1+2\tau^2 & \tau & \tau \\ \tau & 1 & 0 \\ \tau & 0 & 1 \end{bmatrix}.$$

As a result, the components of stress tensor have the form

$$\begin{aligned} &\sigma_{12} = \sigma_{21} = \sigma_{13} = \sigma_{31} = 2C_1(1+\tau)^{-10/3}\tau, \quad \sigma_{32} = \sigma_{23} = 0, \\ &\sigma_{11} = (8/3)C_1(1+\tau)^{-10/3}(\tau-1)\tau + 2D_1\tau(\tau+2), \\ &\sigma_{22} = \sigma_{33} = -(4/3)C_1(1+\tau)^{-10/3}(1+2\tau)\tau + 2D_1\tau(\tau+2). \end{aligned} \tag{30}$$

The formulas (30) show that the Poynting effect (when the values of shear angle increase from the sufficiently small values to the moderate ones, then the shear stress depends nonlinearly on the shear angle) is described by the Neo-Hookean model, because Eq. (30) demonstrates just this nonlinear dependence for the moderate values of shear angle.

Figure 7 shows the dependence of the shear stress on the shear angle $\sigma_{12} \sim \tau$ for the silicon rubber (here and in all next plots, stress is measured by MPa).

5.1.2 Simple shear in model 2

The expressions for gradient **F** and components of tensor **B** are the same as for the Neo-Hookean model. As a result, the expressions from formula (30) are simplified $\lambda_1 = 1$, $\lambda_2 = \lambda_3 = 1+\tau$, $J = (1+\tau)^2$, $I_1 = 1+2\tau^2$, $I_2 = (1+\tau)^2\left[2+(1+\tau)^2\right]$ and components of the stress tensor have the form

$$\sigma_{12} = \sigma_{21} = 2C_{10}(1+\tau)^{-10/3}\tau - 2C_{01}(1+\tau)^{-14/3}(1+4\tau)\tau, \tag{31}$$

$$\sigma_{23} = \sigma_{32} = -2C_{01}(1+\tau)^{-14/3}\tau^2, \tag{32}$$

$$\begin{aligned} \sigma_{11} &= 2C_{10}(1+\tau)^{-10/3}(4/3)\left(1+\tau+2\tau^2\right) + 2D_1\tau(1+2\tau) + \\ &+ 2C_{01}(1+\tau)^{-14/3}(4/3)\left(3+5\tau+5\tau^2+4\tau^3-2\tau^4\right), \end{aligned} \tag{33}$$

$$\begin{aligned} &\sigma_{22} = \sigma_{33} \\ &= 2C_1(1+\tau)^{-2}\left[(1+\tau)^{-4/3} + (1+2\tau^2)\left(1-(1+\tau)^{-4/3}\right) - 1\right] + 2D_1\tau(1+2\tau). \end{aligned} \tag{34}$$

Thus, the Mooney-Rivlin model (that is, more complicated as compared with the Neo-Hookean model) describes the more complicated stress state, which is characterized by six components of the stress tensor. This model describes well-known nonlinear effects. The Poynting effect follows from the representation of the shear stresses by formula (31). The Kelvin effect follows from formulas (33) and (34).

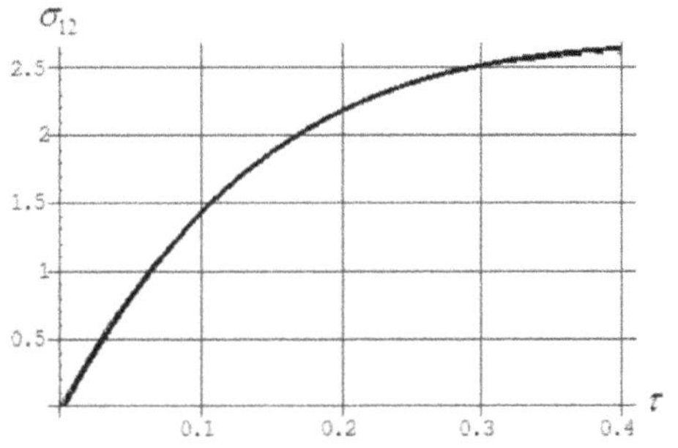

Figure 7.
Dependence of the shear stress on the shear angle $\sigma_{12} \sim \tau$.

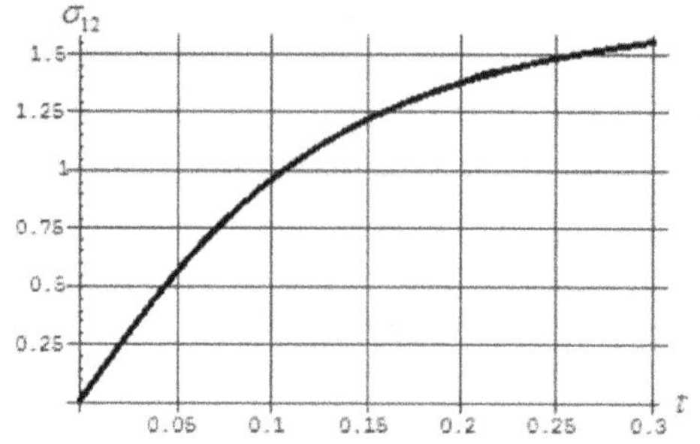

Figure 8.
Dependence of shear stress σ_{12} on the shear strain τ.

Also, formula (32) describes one more nonlinear effect: an initiation of shear stresses $\sigma_{23} = \sigma_{32}$. **Figure 8** shows the nonlinear dependence of shear stress σ_{12} on the shear strain τ, that is built for the silicon rubber. Comparison with **Figure** 7, which corresponds to the Neo-Hookean model, shows that the Mooney-Rivlin model describes the more essential deviation from the linear Hookean description of simple shear.

5.1.3 Simple shear in model 3

The Cauchy-Green strain tensor is characterized by three components

$$\varepsilon_{22} = (1/2)(u_{2,2} + u_{2,2} + u_{k,2}u_{k,2}) = (1/2)\tau^2, \tag{35}$$

$$\varepsilon_{12} = \varepsilon_{21} = (1/2)(u_{1,2} + u_{2,1} + u_{k,1}u_{k,2}) = (1/2)\tau. \tag{36}$$

To calculate the stresses, it is necessary to write the potential (29) concerning the formulas (35) and (36)

$$\begin{aligned} W(\varepsilon_{ik}) = {} & (1/2)\lambda(\varepsilon_{22})^2 + \mu\left[(\varepsilon_{22})^2 + (\varepsilon_{12})^2 + (\varepsilon_{21})^2\right] \\ & +(1/3)A\left[\varepsilon_{22}(\varepsilon_{12}\varepsilon_{12} + \varepsilon_{21}\varepsilon_{21} + \varepsilon_{12}\varepsilon_{21}) + (\varepsilon_{22})^3\right] \\ & +B\left[(\varepsilon_{22})^2 + (\varepsilon_{12})^2 + (\varepsilon_{21})^2\right]\varepsilon_{22} + (1/3)C(\varepsilon_{22})^3, \end{aligned} \tag{37}$$

$$W(\tau) = (1/2)\mu\tau^2 + (1/8)[(\lambda + 2\mu) + A + B]\tau^4 + (1/24)[A + 3B + C]\tau^6 \tag{38}$$

The Lagrange stress tensor is determined by the formula $\sigma_{ik}(x_n, t) = \partial W / \partial \varepsilon_{ik}$ and has two nonlinear com-ponents

$$\begin{aligned} \sigma_{22} = {} & (\lambda + 2\mu)\varepsilon_{22} + A\left[(\varepsilon_{22})^2 + (1/3)(\varepsilon_{12}\varepsilon_{12} + \varepsilon_{21}\varepsilon_{21} + \varepsilon_{12}\varepsilon_{21})\right] \\ & +B\left[3(\varepsilon_{22})^2 + (\varepsilon_{12})^2 + (\varepsilon_{21})^2\right] + C(\varepsilon_{22})^2 \\ = {} & (1/4)[2(\lambda + 2\mu) + (A + 2B)]\tau^2 + (1/4)(A + 3B + C)\tau^4, \end{aligned} \tag{39}$$

$$\sigma_{12} = \sigma_{21} = 2\mu\varepsilon_{12} + [(1/3)A(\varepsilon_{12} + \varepsilon_{21}) + 2B\varepsilon_{12}]\varepsilon_{22} = \mu\tau + (1/6)(A + 3B)\tau^3. \tag{40}$$

The shear stress contains the linear and nonlinear summands and describes the simple shear. The normal stress describes the change of volume under deformation and testifies the break of the state of simple shear in the nonlinear description of deformation. To build the plots of dependence (40) choose two nonstandard for the Murnaghan model materials – foam and polystyrene – which can experience not only the small by values strains but also the moderate ones. **Figures 9** and **10** show the dependence of the shear stress σ_{12} on the shear angle τ for the foam and polystyrene.

The dependences $\sigma_{12} \sim \tau$ for models 1–3 can be commented in the following way: these models describe well the nonlinear Poynting effect. At the same time, many scientists working with auxetic materials report the experimental dependences that coincide quantitatively with the shown here theoretical (and based on them numerical) dependences (for example, [46–48]). Also, some conclusions to dependence $\sigma_{12} \sim \tau$ for models 1–3 can be formulated: the developed in mechanics of materials nonlinear models of deformation of elastic materials can be recommended for the description of auxetic materials.

5.2 Universal deformation of uniaxial tension

This kind of deformation is fundamental for the auxetics because just in tests on the uniaxial tension-compression the phenomenon of auxeticity was first observed.

5.2.1 Uniaxial tension in model 1

The formulas for the principal extensions $\lambda_2 = \lambda_3, J = \lambda_1\lambda_2^2, I_1 = \lambda_1^2 + 2\lambda_2^2$ are valid in this model and the normal stresses (the shear stresses are absent in this state of deformation) are given by the formulas

$$\sigma_{11} = (2/3)\mu\left(\lambda_1\lambda_2^2\right)^{-5/3}\left(\lambda_1^2 - \lambda_2^2\right) + k\left(\lambda_1\lambda_2^2 - 1\right), \tag{41}$$

$$\sigma_{22} = \sigma_{33} = -(1/3)\mu\left(\lambda_1\lambda_2^2\right)^{-5/3}\left(\lambda_1^2 - \lambda_2^2\right) + k\left(\lambda_1\lambda_2^2 - 1\right). \tag{42}$$

Note that the stresses are depending in model 1 on two principal extensions – longitudinal and transverse.

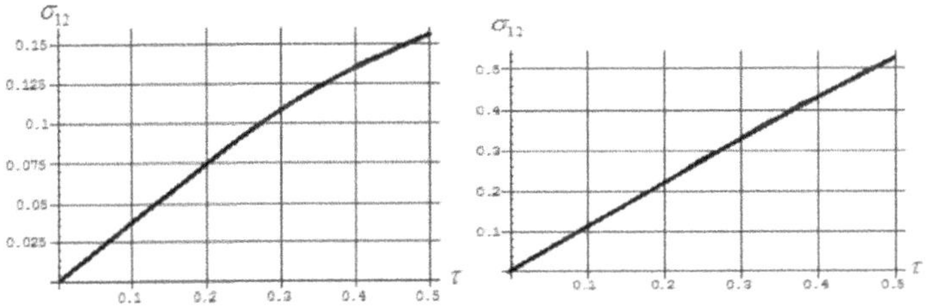

Figure 9.
Dependence of the shear stress on the shear angle (foam).

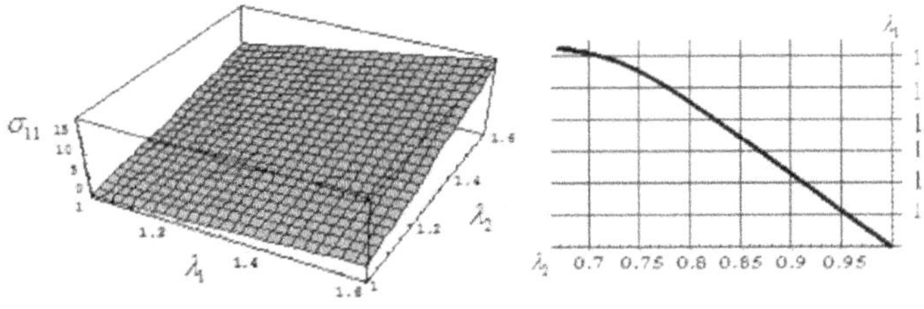

Figure 10.
Dependence of the shear stress on the shear angle (polystyrene).

If to assume that all three normal stresses on the lateral surface of the sample are absent (the surface is free of stresses), then

$$\sigma_{11} = 3k\left(\lambda_1\lambda_2^2 - 1\right). \tag{43}$$

It follows from (43) that the Poynting-type effect (when the principal extensions increase from the sufficiently small values to the moderate ones, then the normal stress in the direction of tension depends nonlinearly on these extensions) is described by the Neo-Hookean model.

Figure 11 shows the dependence of the longitudinal stress on principal extensions and is built for the rubber with allowance for that the value $(\mu/3k) = 0,00334$ is very small compared to the unit (the bulk mo-dulus is essentially more of the shear one). Then the dependence is valid

$$\varepsilon_{22} = (1/2)(1 + 2\varepsilon_{11})^{-2} - 1/2. \tag{44}$$

Figure 12 corresponds to formulas (41) and (42). and shows a dependence of the longitudinal principal extension on the transverse principal extension. Note that the silicon rubber is characterized by the big difference between values of shear and bulk moduli that can reach hundred times. Therefore, the new material is chosen further for the numerical analysis – the foam, which values of elastic constants is characterized by about equal by the order. **Figure 12** shows also that with an increase of extension λ_1 the increase of extension λ_2 slows.

It looks, in this case, to be illogical to neglect the first summand in (41) and (42). Note here that the ratio (λ_2/λ_1) corresponds in the linear theory to the Poisson's ratio.

5.2.2 Uniaxial tension in model 2

The uniaxial tension in direction of the abscissa axis is characterized by parameters: $\lambda_2 = \lambda_3, J = \lambda_1\lambda_2^2, I_1 = \lambda_1^2 + 2\lambda_2^2, I_2 = \lambda_2^4 + 2\lambda_1^2\lambda_2^2, B_{11} = \lambda_1^2, (BB)_{11} = \lambda_1^4$. The normal stresses are given by the formulas

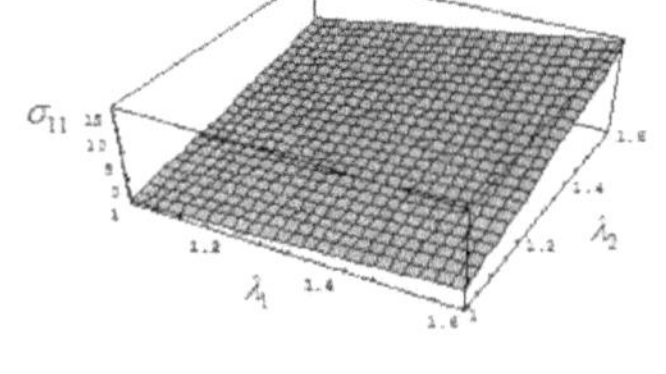

Figure 11.
Dependence of the longitudinal stress on the principal extensions.

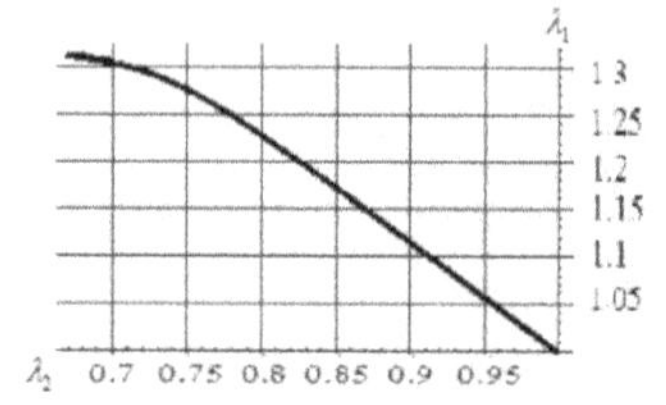

Figure 12.
Dependence of the longitudinal principal extension on the transverse principal extension.

$$\sigma_{11} = 2C_{10}(2/3)\left(\lambda_1\lambda_2^2\right)^{-5/3}\left(\lambda_1^2 - \lambda_2^2\right) + 2C_{01}\left(\lambda_1\lambda_2^2\right)^{-7/3}\left(\lambda_1^4 + (2/3)\lambda_1^2\lambda_2^2 - (5/3)\lambda_2^4\right)$$
$$+2D_1\left(\lambda_1\lambda_2^2 - 1\right), \tag{45}$$

$$\sigma_{22} = \sigma_{33} = (2/3)C_{10}\left(\lambda_1\lambda_2^2\right)^{-5/3}\left(\lambda_2^2 - \lambda_1^2\right) + 2C_{01}\left(\lambda_1\lambda_2^2\right)^{-7/3}(1/3)\lambda_2^2\left(\lambda_2^2 - \lambda_1^2\right) + 2D_1\left(\lambda_1\lambda_2^2 - 1\right). \tag{46}$$

Assume that all three normal stresses over the sample lateral surface are absent. Then Eq. (45) is simplified to the form

$$\sigma_{11} = 2C_{01}\left(\lambda_1\lambda_2^2\right)^{-7/3}\left(\lambda_1^4 - \lambda_2^4\right) + 6D_1\left(\lambda_1\lambda_2^2 - 1\right). \tag{47}$$

The last formula testifies: the Mooney-Rivlin model describes the Poynting-type effect.

Two elastic constants are presented in (47) in contrast to the Neo-Hookean model, where the shear modulus was absent. It should be noted that in both models – Neo -Hookean and Moo-ney-Rivlin –the tension in the longitudinal direction stress σ_{11} depends already on two principal extensions. **Figure 13** shows a dependence of the longitudinal stress on principal extensions is built for the silicon rubber. It coincides practically with **Figure 11** (Neo-Hookean model) and shows that the constant C_{01} of the Mooney-Rivlin model effects not essentially on the stress σ_{11} and the dependence (45) rests weakly nonlinear within the accepted restrictions.

The Eq. (46) can be transformed into the form

$$\lambda_1^6 - \lambda_1^3/\sigma^2 + \left[(2C_{10}/6D_1)\sqrt[-3]{\sigma^4} + (2C_{01}/6D_1)\sqrt[-3]{\sigma^2}\right]\sigma^{-4}\left(\sigma^2 - 1\right) = 0,$$
$$\sigma = (\lambda_2/\lambda_1),\quad \lambda_1^3 = 1/2\sigma^2 \pm 1/2\sigma^2\sqrt{1 - \begin{bmatrix}(2C_{10}/6D_1)\sqrt[-3]{\sigma^4} \\ +(2C_{01}/6D_1)\sqrt[-3]{\sigma^2}\end{bmatrix}\sigma^{-4}\left(\sigma^2 - 1\right)}, \tag{48}$$

The corresponding to the model 1 plot from **Figure 11** is practically identical with the plot from **Figure 13** corresponding to model 2.

5.2.3 Uniaxial tension in model 3

The uniaxial tension in this model is characterized by three nonzero components of the strain tensor ε_{kk} and one non-zero component of the stress tensor σ_{11}. Then the constitutive equations are somewhat simp-lified.

$$\sigma_{11} = \lambda I_1 + 2\mu\varepsilon_{11} + A(\varepsilon_{11})^2 + B(\mathrm{E} + 2\varepsilon_{11}I_1) + C(\mathrm{E} + 2\varepsilon_{22}\varepsilon_{11} + 2\varepsilon_{33}\varepsilon_{11})$$
$$I_1 = \varepsilon_{11} + \varepsilon_{22} + \varepsilon_{33},\quad \mathrm{E} = (\varepsilon_{11})^2 + (\varepsilon_{22})^2 + (\varepsilon_{33})^2. \tag{49}$$

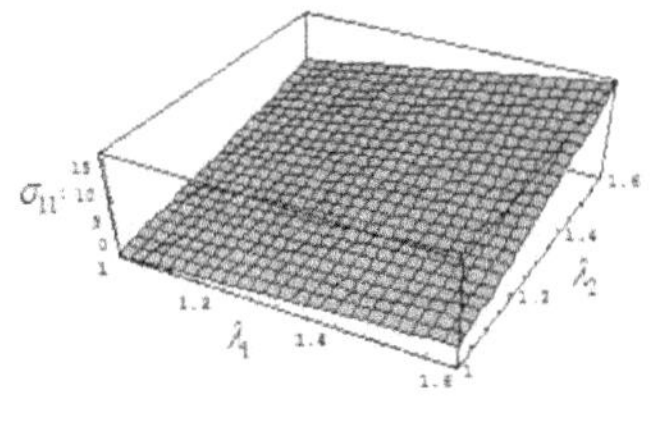

Figure 13.
Dependence of the longitudinal stress on principal extensions.

$$0=\lambda I_1+2\mu\varepsilon_{22}+A(\varepsilon_{22})^2+B(\mathrm{E}+2\varepsilon_{22}I_1)+C(\mathrm{E}+2\varepsilon_{22}\varepsilon_{33}+2\varepsilon_{22}\varepsilon_{11}), \tag{50}$$

$$0=\lambda I_1+2\mu\varepsilon_{33}+A(\varepsilon_{33})^2+B(\mathrm{E}+2\varepsilon_{33}I_1)+C(\mathrm{E}+2\varepsilon_{22}\varepsilon_{11}+2\varepsilon_{22}\varepsilon_{33}). \tag{51}$$

Let us remind that in the linear theory of elasticity, corresponding to the Hookean model, the constitutive equations are significantly simpler

$$\sigma_{11}=\lambda I_1+2\mu\varepsilon_{11},\ 0=\lambda I_1+2\mu\varepsilon_{22},\ 0=\lambda I_1+2\mu\varepsilon_{33}. \tag{52}$$

Apply further to the nonlinear Eqs. (49)–(51) the procedure of analysis of the state of uniaxial tension that is used in the linear theory of elasticity as applied to Eqs. (52). Subtraction of Eq. (51) from Eq. (50) gives the formula

$$0=2\mu(\varepsilon_{22}-\varepsilon_{33})+A\left((\varepsilon_{22})^2-(\varepsilon_{33})^2\right)+2B(\varepsilon_{22}-\varepsilon_{33})(\varepsilon_{11}+\varepsilon_{22}+\varepsilon_{33}),$$

from which the equality of components of transverse strains $\varepsilon_{22}=\varepsilon_{33}$ follows. The addition of formulas (36)–(38) results in the following formula

$$\begin{aligned}&\sigma_{11}/(3\lambda+2\mu)-[(A+3B+C)/3\lambda+2\mu]\left[(\varepsilon_{11})^2+2(\varepsilon_{22})^2\right]-[2B/(3\lambda+2\mu)](\varepsilon_{11}+2\varepsilon_{22})^2\\&-[4C/(3\lambda+2\mu)]\left[(\varepsilon_{22})^2+2\varepsilon_{22}\varepsilon_{11}\right]=\varepsilon_{11}+2\varepsilon_{22}.\end{aligned} \tag{53}$$

Substitution of formula (53) into the relation (49) gives new relation

$$\begin{aligned}\sigma_{11}=E\varepsilon_{11}&+\left(A+\frac{2\lambda+3\mu}{\lambda+\mu}B+C\right)(\varepsilon_{11})^2-\frac{\lambda}{\lambda+\mu}\left(A+\frac{4\lambda-2\mu}{\lambda}B-\frac{2\mu}{\lambda}C\right)(\varepsilon_{22})^2\\&+\frac{2(\lambda+2\mu)}{\lambda+\mu}(B+C)\varepsilon_{11}\varepsilon_{22}.\end{aligned} \tag{54}$$

The relation (54) shows that model 3, like models 1 and 2, describes the Poynting-type effect.

Figures 14 and **15** show the dependence $\sigma_{11}=\sigma_{11}(\varepsilon_{11},\varepsilon_{22})$ among the longitudinal stress σ_{11} and strains $\varepsilon_{11},\varepsilon_{22}$ for the foam and polystyrene and the moderate values of strains. Both plots demonstrate an essential nonlinearity under moderate strains. This new nonlinear effect will be true for auxetic materials too.

Write now the constitutive Eq. (49) with allowance for equality $\varepsilon_{22}=\varepsilon_{33}$ and transform it to the form of a quadratic equation relative to the ratio $\varepsilon_{22}/\varepsilon_{11}$

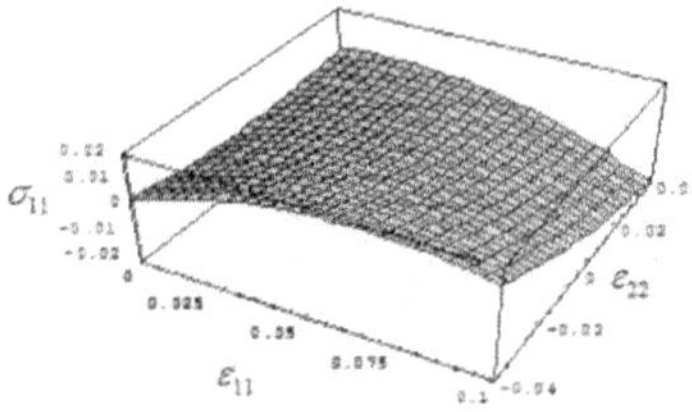

Figure 14.
Dependence $\sigma_{11}=\sigma_{11}(\varepsilon_{11},\varepsilon_{22})$ *(foam).*

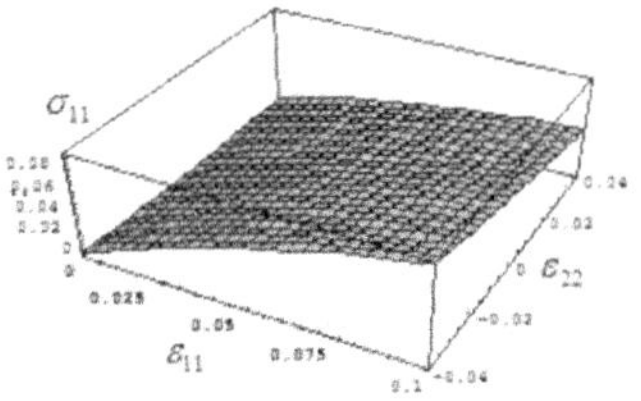

Figure 15.
Dependence $\sigma_{11} = \sigma_{11}(\varepsilon_{11}, \varepsilon_{22})$ *(polystyrene).*

$$\left(\frac{\varepsilon_{22}}{\varepsilon_{11}}\right)^2 + 2\frac{[((\lambda+\mu)/\varepsilon_{11}) + (B+C)]}{(A+6B+4C)}\frac{\varepsilon_{22}}{\varepsilon_{11}} + \frac{(\lambda/\varepsilon_{11}) + (B+C)}{(A+6B+4C)} = 0.$$

The solution of this equation has the form

$$(\varepsilon_{22}/\varepsilon_{11}) = -\{[(\lambda+\mu)/\varepsilon_{11} + (B+C)]/(A+6B+4C)\}\left[1 \pm \sqrt{1 - \frac{(A+6B+4C)[\lambda/\varepsilon_{11} + (B+C)]}{[(\lambda+\mu)/\varepsilon_{11} + (B+C)]^2}}\right]. \tag{55}$$

Thus, Eq. (55) shows that the ratio $(-\varepsilon_{22}/\varepsilon_{11})$ is not constant in the Murnaghan nonlinear model. This can be treated as the new mechanical nonlinear effect which is looking very promising for the auxetic materials.

Figures 16 and **17** show a dependence of the ratio $(-\varepsilon_{22}/\varepsilon_{11})$ on the strain ε_{11} and are built for the foam and polystyrene for the moderate strains. The plot's main features are as follows: the ratio $(-\varepsilon_{22}/\varepsilon_{11})$ is de-creased essentially from the initial value, which corresponds to the Poisson ratio for small strain in the con-ventional materials, to the negative values under the moderate values of longitudinal strain that is observed in the auxetic materials. So, the ratio, that is, treated as the Poisson's ratio for small strain, in the case of mo-derate strain becomes the characteristics of transition of the material from the category of conventional ma-terials into the

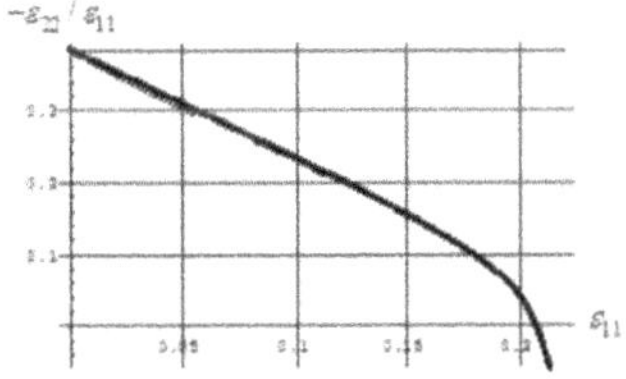

Figure 16.
Dependence of the ratio $(-\varepsilon_{22}/\varepsilon_{11})$ *on the strain* ε_{11} *(foam).*

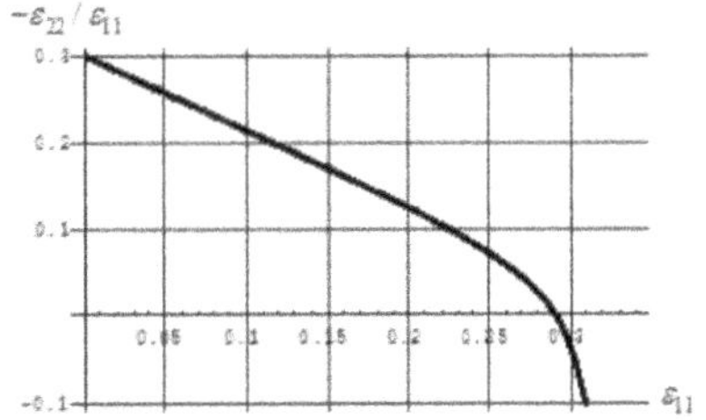

Figure 17.
Dependence of the ratio $(-\varepsilon_{22}/\varepsilon_{11})$ *on the strain* ε_{11}*(polystyrene).*

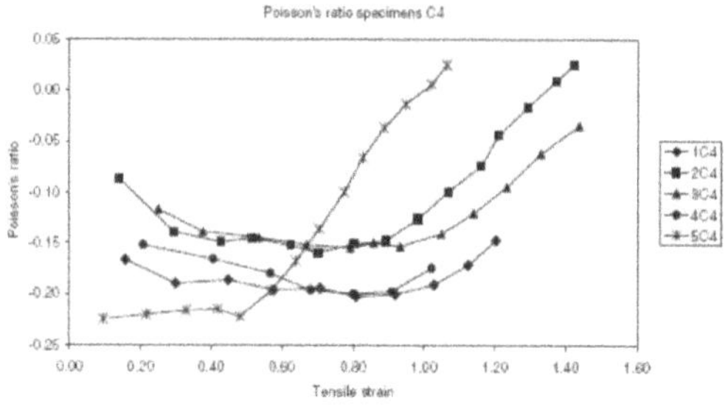

Figure 18.
Experimental dependence of the ratio $(-\varepsilon_{22}/\varepsilon_{11})$ on the strain ε_{11}.

category of nonconventional materials. This can be considered as the newly revealed theore-tically nonlinear effect.

Thus, an analysis of universal deformation of uniaxial tension for model 3 revealed the new property: the material with conventional properties under small strains is transformed under moderate strains into the nonconventional (auxetic) material. The uncommonness of this observation consists in that usually the material is considered either the conventional or the nonconventional during all the processes of deformation.

Let us compare the plots from **Figures 16** and **17** with the experimental data from ([49], Figure 4) shown here as **Figure 18** (dependence of the ratio $(-\varepsilon_{22}/\varepsilon_{11})$ on the strain ε_{11}), where the deformation of the foams was studied for the finite strains with increasing the longitudinal strain ε_{11} from 0.1 to 1.4. Note that the theoretical plots are constructed for the range from $\varepsilon_{11} = 0$ to the moderate values 0.23 (foam) and 0.33 (polystyrene). This comparison shows that $(\varepsilon_{22}/\varepsilon_{11})$ increases within the range $\varepsilon_{11} \in (0,0;0,3)$. Thus, model 3 describes some experimental observations of the foam.

Figures 19 and **20** show the dependence of longitudinal and transverse strains. Three stages can be marked out: 1. A decrease of transverse strain becomes slower under transition to the moderate strains. 2. The strain ε_{22} reaches the local minimum

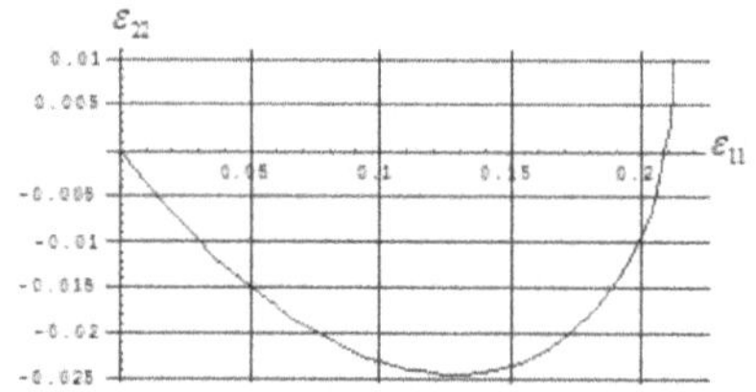

Figure 19.
Dependence of longitudinal and transverse strains (foam).

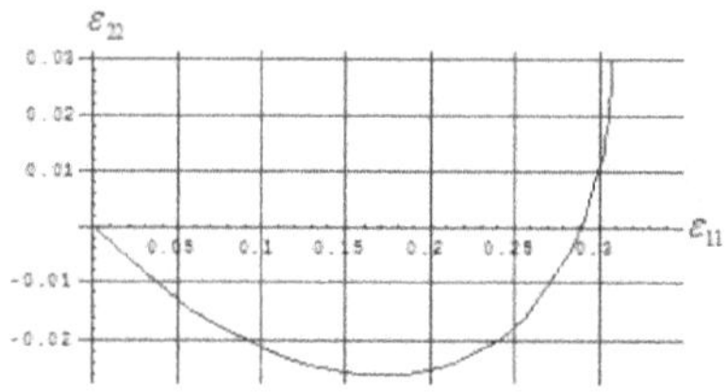

Figure 20.
Dependence of longitudinal and transverse strains (polystyrene).

and further increases. 3. When the strain ε_{11} continues to increase, the strain ε_{22} possesses zero value and further increases possessing already positive values.

The shown feature confirms once again the new mechanical effect – a transition of the material under its deformation to the level of moderate values of the longitudinal stretching from the class of conventional ma-terials into the class of the auxetic materials. In other words, the standard sample in conditions of universal deformation of uniaxial tension is deformed for small strains as if it is made of the conventional material (its cross-section is decreased) and with increasing the values of longitudinal stretching to the moderate values the sample cross-section starts to increase, what is the characteristic just for auxetic materials.

The plots from **Figures 19** and **20** can be compared with the plot, obtained experimentally in [48]. This article reports that the new metamaterials were created from soft silicon rubber. The samples we-re deformed in conditions of uniaxial compression up to moderate values of longitudinal strain 0,35. The shown in the **Figure 21** plot corresponds to **Figure 2a** in [48] and shows a dependence of longitudinal and transverse strains. Comparison of plots from **Figure 11** (uniaxial stretching) and **Figure 12** (uniaxial compression) demonstrates the common property of forming the hump in the area of negative values of transverse strain, which is transformed with the increasing values of longitudinal strain roughly into the straight line in the area of positive values of transverse strain.

Thus, the nonlinear Murnaghan model describes within conditions of uniaxial tension some nonlinear phenomena of deformation, which can be linked with the properties of deformation of auxetic materials. Note that the shown feature is clearly visible only within the framework of the Murnaghan model, but the Neo-Hookean and Mooney -Rivlin models also describe the hump formation, as can be seen in **Figure 6**.

5.3 Universal deformation of omniaxial tension

5.3.1 Omniaxial tension in model 1

In this case $\lambda_1 = \lambda_2 = \lambda_3, J = \lambda_1^3,\ I_1 = 3\lambda_1^2$ and the normal stress is equal

$$\sigma_{11} = 2D_1\left(\lambda_1^3 - 1\right). \tag{56}$$

The formula (56) describes the Poynting-type effect relative to the bulk modulus (the dependence σ_{11} on the extension λ_1 is evidently nonlinear).

Figure 22 shows a dependence of the stress on the principal extension and is built for the silicon rubber. The plot testifies that model 1 describes the nonlinear

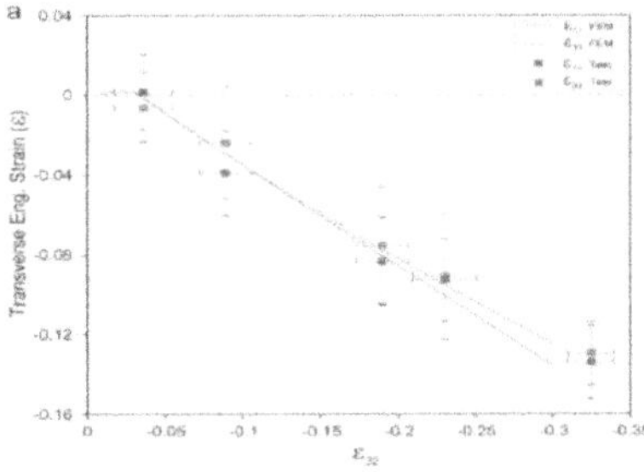

Figure 21.
Experimental dependence of longitudinal and transverse strains.

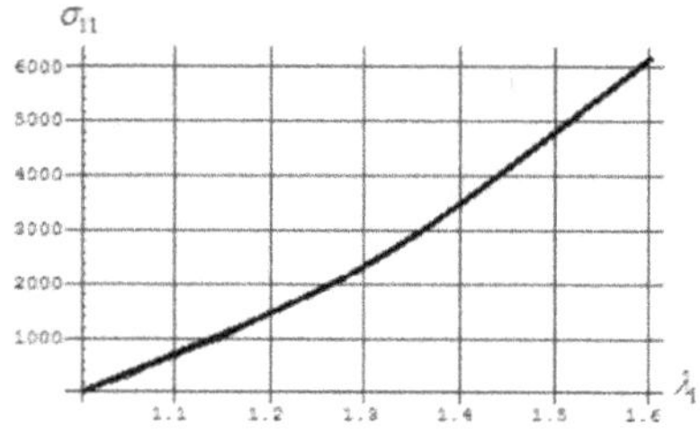

Figure 22.
Dependence of stress on principal extension.

change of the sample volume while being subjected to the universal deformation of uniform compression-tension.

5.3.2 Omniaxial tension in model 2

In this case $\lambda_1 = \lambda_2 = \lambda_3, J = \lambda_1^3,\ I_1 = 3\lambda_1^2, I_2 = 3\lambda_1^4$ (note that they are identical for any nonlinear model). The formula for normal stress coincides with the analogous formula for model 1 (56) and verifies the non-linear dependence of the tension stress on the principal extension.

5.3.3 Omniaxial tension in model 3

The components of displacement gradients and Cauchy-Green strain tensor are as follows

$$\begin{aligned} u_{1,1} &= u_{2,2} = u_{3,3} = \varepsilon > 0, u_{1,1} + u_{2,2} + u_{3,3} = 3\varepsilon = e, u_{k,m} = (\partial u_k/\partial x_m) \\ &= 0 \quad (k \neq m); \varepsilon_{11} = \varepsilon_{22} = \varepsilon_{33} = \varepsilon + (1/2)\varepsilon^2, \quad \varepsilon_{ik} = 0 \quad (i \neq k) \end{aligned} \tag{57}$$

The corresponding algebraic invariants of the Cauchy-Green tensor are written in the form

$$\begin{aligned} I_1 &= \varepsilon_{11} + \varepsilon_{22} + \varepsilon_{33} = e, I_2 = (\varepsilon_{11})^2 + (\varepsilon_{22})^2 + (\varepsilon_{33})^2 = (1/3)e^2, I_3 \\ &= (\varepsilon_{11})^3 + (\varepsilon_{22})^3 + (\varepsilon_{33})^3 = (1/9)e^3. \end{aligned} \tag{58}$$

The formulas for invariants (58) allow writing the potential in the simpler form

$$\begin{aligned} W(\varepsilon) &= (3/2)(3\lambda + 2\mu)\varepsilon^2 + ((9/2)\lambda + 3\mu + A + 9B + 9C)\varepsilon^3 + \\ &+ (3/2)(4(3\lambda + 2\mu) + (A + 9B + 9C))\varepsilon^4 + (3/4)(A + 9B + 9C)\varepsilon^5 + (1/8)(A + 9B + 9C)\varepsilon^6. \end{aligned} \tag{59}$$

The stresses are evaluated by the formulas (the normal stresses only are nonzero)

$$\begin{aligned} &\sigma_{11} = \sigma_{22} = \sigma_{33} = (3\lambda + 2\mu)\varepsilon + [(3/2)(3\lambda + 2\mu) + (A + 9B + 7C)]\ \varepsilon^2 \\ &+ (A + 9B + 7C)\left(\varepsilon^3 + (1/4)\varepsilon^4\right), \quad \sigma_{12} = \sigma_{23} = \sigma_{31} = 0. \end{aligned}$$

Thus, the stresses contain linear and nonlinear summands.

The interdependence between the first invariant of the stress tensor σ_{kk} and the parameter of the omni-axial tension e has the form

$$\sigma_{kk} = (3\lambda + 2\mu)e + [(1/2)(3\lambda + 2\mu) + (1/3)(A + 9B + 7C)]e^2 \\ +(A + 9B + 7C)[(1/9)e^3 + (1/108)e^4]. \tag{60}$$

The plots in **Figures 23** and **24** show a dependence $\sigma_{kk}(e)$ for the foam and polystyrene evaluated formula (60). It follows from them that they are similar to the parabola with a vertex in a positive half of the plane $\sigma_{kk}Oe$. The parabola's right branch then passes into the negative half of the plane. Both plots have "the hump" in the positive branch of the plane.

A presence of "the hump" testifies that the nonlinear Murnaghan model describes the transition of the material of the sample-cube from the class of conventional materials into the class of auxetic materials. The fact is that the sample is compressed for the small values of uniform tension and in the following increase of the tension the strain the sample swells. But this phenomenon is characteristic of only auxetic materials.

Thus, three nonlinear models which are used in the analysis describe the nonlinear Poynting-type effects in conditions of three used above universal deformations and the moderate strains. This agrees quantitatively with experimental observations of nonlinear dependences $\sigma \sim \varepsilon$ (stress versus strain) in auxetic materials for the moderate strains.

The main new effects are revealed: the nonlinear Murnaghan model describes in the case of uniaxial and omniaxial tension the transition of the material from the class of conventional materials into the class of the auxetic materials. This occurs when the material is deformed to the level of moderate values of the longitudinal stretching. In other words, the shown experiments and proposed theoretical analysis testify that the stan-dard sample in conditions of the mentioned universal deformation of uniaxial tension is deformed for small strains as if it is made of the conventional material (its cross-section is decreased) and with increasing the values of longitudinal stretching to the moderate values the sample cross-section starts to increase, what is the characteristic just for auxetic materials.

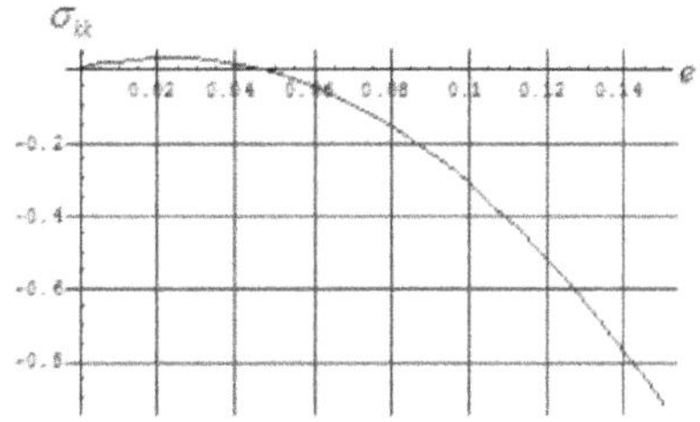

Figure 23.
Dependence $\sigma_{kk}(e)$ (foam).

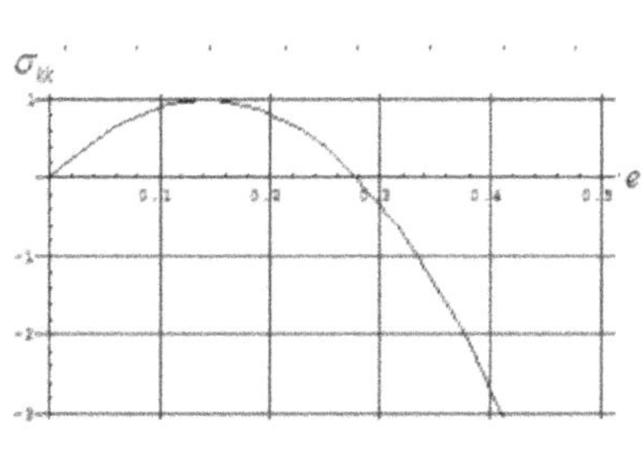

Figure 24.
Dependence $\sigma_{kk}(e)$ (polystyrene).

6. Final conclusions

The elasticity is the property of auxetic materials, which is especially characteristic and most studied for these materials. Historically, the auxetics were treated from the point of view of the linear theory of elasticity what was not quite adequate in some cases.

As the part of classical mechanics of elastic materials, the mechanics of auxetic materials needs at present more and more experimental studies (the level of such studies as compared with the classical ones can be seen from the famous Bell's book [27]).

The nonlinear theory of elasticity is seemingly quite prospective for a description of elastic deformation of the auxetic materials but it is essentially more complicated in the mathematical apparatus and concrete investigations.

Author details

Jeremiah Rushchitsky
S.P. Timoshenko Institute of Mechanics, Kyiv, Ukraine

*Address all correspondence to: rushch@inmech.kiev.ua

References

[1] Lim T.C. Auxetic materials and structures. Berlin: Springer; 2015. 564 p.

[2] Lim T.C. Mechanics of Metamaterials with Negative Parameters. Singapore: Springer; 2020. 712 p.

[3] Hu H. Zhang M. Liu Y. Auxetic Textiles. Cambridge: Woodhead Publishing; 2019. 366 p.

[4] Alderson A., Alderson K.L. Auxetic materials. *I. Mech*. E.: J. Aerosp. Eng. 2007; **221**. N4: 565-575.

[5] An introduction to auxetic materials: an interview with Professor Andrew Alderson, AZoMaterials, August 29, 2015.

[6] Cabras L, Brun M. Auxetic two-dimensional lattices with Poisson's ratio arbitrarily close to −1 Proc. Roy. Soc. London A. 2014;**470**(20140538):1-23

[7] Carneiro VH, Meireles J, Puga H. Auxetic Materials – A Review. Materials Science – Poland. 2013;**31** (N4):561-571

[8] Gibson L.J., Ashby M.F. Cellular Solids: Structure and Properties. Cambridge: Cambridge University Press, 1999. 2nd ed. 2010. –510p.

[9] Greaves GN. Poisson's ratio over two centuries: challenging hypotheses. Notes and Records of Roy. Soc. 2013;**67**(N1): 37-58

[10] Grima JN. Auxetic metamaterials. Strasbourg, France: European Summer Campus; 2010. pp. 1-13 http://www.auxetic.info

[11] Hou X, Silberschmidt VV. Metamaterials with negative poisson's ratio: a review of mechanical properties and deformation mechanisms. In: *Mechanics of Advanced Materials*. Berlin: Springer; 2015. pp. 155-179

[12] Kolken HMA, Zadpoor AA. Auxetic mechanical metamaterials. The Royal Society of Chemistry Advances. 2017; **N7**:511-5130

[13] Liu YP, Hu H. A review on auxetic structures and polymeric materials. Scientific Research and Essays. 2010;**5** (N10):1052-1063 http://www.auxetic.info

[14] Materials. Special issue "Auxetics 2017-2018".

[15] Pravoto Y. Seeing auxetic materials from the mech anics point of view: A structural review on the negative Poisson's ratio. Computational Materials Sci. 2012;**58**:140-153

[16] Saxena KK, Das R. Calius E. P. Advanced Engineering Materials. 2016

[17] Scarpa F. Auxetics: From Foams to Composites and Beyond (presentation to Sheffield May 2011) http://www.bris.ac.uk/composites

[18] Scarpa F, Pastorino P, Garelli A, Patsias S, Ruzzene M. Auxetic compliant flexible PU foams: static and dynamic properties. Physica Status Solidi B. 2005; **242**(N3):681-694

[19] Gibson LJ, Ashby MF, Schayer GS, Robertson CI. The Mechanics of Two-Dimensional Cellular Materials. Proc. Roy. Soc. Lond. A. 1982; **382**:25-42

[20] Gibson LJ, Ashby MF. The Mechanics of Three-Dimensional Cellular Materials. Proc. Roy. Soc. Lond. A. 1982;**382**:43-59

[21] Evans KE. Auxetic polymers: a new range of materials. Endeavour. 1991;**15**: 170-174

[22] Anurag C, Anvesh CK, Katam S. Auxetic materials. Int. J. for Research in

Appl. Science & Eng. Techno logy. 2015; **3**(N4):1176-1183

[23] Dudek K.K., Attard D., Caruana-Gauci R., Wojciecho wski K.W., Grima J.N. Unimode metamaterials exhibiting negative linear compressibility and negative thermal expansion. Smart Materials and Structures. 2016; **25**, N2: 025009.

[24] Lakes RS. Foam Structures with a Negative Poisson's Ratio. Science. 1987; **235**:1038-1040

[25] Wojciechowski KW. Constant thermodynamic tension Monte-Carlo studies of elastic properties of a two-dimensional system of hard cyclic hexameters. Molecular Physics. 1987;**61**: 1247-1258

[26] Truesdell C. A first course in rational continuum mechanics. Baltimore: The John Hopkins University; 1972; 2nd ed., New York: Academic Press; 1991. 352 p.

[27] Flügge's Encyclopedia of Physics. Vol. VIa/I. Bell J.F. Mechanics of solids. Berlin: Springer-Verlag; 1973. 468 p.

[28] Hetnarski R.B., Ignaczak J. The mathematical theory of elasticity. Boca Raton: CRC Press; 2011. 804 p.

[29] Lurie A.I. Theory of elasticity. Berlin: Springer; 2005. 1050 p.

[30] Rushchitsky JJ. On universal deformations in an ana lysis of the nonlinear Signorini theory for hyperelastic me-dium. Int. Appl. Mech. 2007;**43**(N12):1347-1350

[31] Rushchitsky J.J. Nonlinear elastic waves in materials. Heidelberg: Springer; 2014. 454 p.

[32] Mooney M. A theory of large elasic deformations. J. Appl. Phys. 1940;**11** (N9):582-592

[33] Rivlin RS. Large elastic deformations of isotropic ma-terials. IV. Further development of general theory Phil. Trans. Roy. Sci. London, Ser. A. Math. and Phys. Sci. 1948;**241(835)**: 379-397

[34] Lur'e A.I. Nonlinear theory of elasticity. Amsterdam: North-Holland; 1990. 617 p.

[35] Ogden R.W. Nonlinear elastic deformations. New York: Dover;1997. 544 p.

[36] Love A.E.H. The mathematical theory of elasticity. 4th ed. New York: Dover; 1944. 643 p.

[37] Germain P. Cours de mechanique des milieux continus. Tome 1. Theorie generale. Paris: Masson et Cie, Editeurs; 1973. 270 p.

[38] Nowacki W. Teoria sprężystośći. Warszawa: PWN; 1970. 769 p (In Polish)

[39] Hahn HG. Elastizitätstheorie. Grundlagen der linea-ren Theorie und Anwendungen auf eindimensionale, ebene und räumliche Probleme. Stuttgart: Teubner; 1985. 336 p (In German)

[40] Leibensohn LS. Short course of the theory of elasticity. Gostechizdat: Moscow-Leningrad; 1942. 304 p (In Russian)

[41] Landau L.D., Lifshits E.M. Theoretical physics. In 10 vols. Vol. VII. Theory of Elasticity. Oxford: Pergamon Press; 1970. 204 p.

[42] Holzapfel G.A. Nonlinear solid mechanics. A conti nuum approach for engineering, – Chichester: Wiley, 2006. 470 p.

[43] Guz A.N. Elastic waves in bodies with initial stresses. In 2 vols. V.1. General problems. V.2. Regularities of

pro-pagation. Kyiv: Naukova Dumka, 1986. 376 p., 536 p. (In Russian)

[44] Murnaghan F.D. Finite Deformation in an Elastic Solid. New York: John Wiley; 1951(1967). 218 p.

[45] Hauk V. (ed) Structural and residual stress analysis. – Amsterdam: Elsevier, 1997 (e-variant 2006). 640 p.

[46] Uzun M. Mechanical properties of auxetic and con-ventional polyprophylene random short fibre reinforced composites. Fibres and Textiles in Eastern Europe. 2012; **20,** N5 (94): 70-74.

[47] Yao YT, Uzun M, Patel I. Working of auxetic nano -materials. J. Achievements in Materials and Manufactur-ing Engineering. 2011;**49** (N2):585-594

[48] Dagdelen J, Montoya J, de Jong M, Persson K. Computational prediction of new auxetic materials. Nature. Communications. 2017;**323**:1-8

[49] Babaee S, Shim J, Weaver JC, Chen ER, Patel N, Bertoldi K. 3D Soft Metamaterials with Negative Poisson's Ratio. Advanced Materials. 2013. DOI: 10.1002/adma.201301946. www. advmat.de

Chapter 7

Perspective Chapter: Improvement of Elastomer Elongation and Output for Dielectric Elastomers

Seiki Chiba, Mikio Waki, Shijie Zhu, Tonghuan Qu and Kazuhiro Ohyama

Abstract

The need for light, high-strength, and artificial muscles is growing rapidly. A well-known type of artificial muscle meeting these requirements is the dielectric elastic (DE) type, which uses electrostatic force between electrodes. In hopes of utilizing, it practically for a variety of purposes, research and development is rapidly progressing all over the world as a technology for practical use. Much of the market demand is dominated by more output-focused applications such as DE power suits, DE motors, DE muscles for robots, and larger DE power systems. To meet these demands, the elasticity of the elastomer is very important. In this paper, we discussed what the important factors are for SS curves, viscoelasticity tests, etc. of the dielectric elastomer materials. Recent attempts have been also made to use new carbon foam materials such as SWCNTs and MWCNTs as electrodes for DEs. These electrodes bring the elastomers to a higher level of performance.

Keywords: Dielectric elastomer, Actuator, Sensor, Generator, Large deformation, High efficiency, Artificial Muscle

1. Introduction

The creation of artificial muscle has long been a scientific aspiration. It is well known that Wilhelm Conrad Röntgen, who discovered X-rays, conducted experiments using rubber strings as artificial muscles [1]. In the 1950s, artificial muscles using EPA (Electro Active Polymer) became mainstream. Since then, the need for light, high-strength, and artificial muscles has been growing rapidly.

EAP type artificial muscles which drive a polymer membrane by applying electrical stimulation, are actuators that realize movements similar to living muscles by electrical control. Because they move softly, they are also called soft actuators. **Figure 1** shows the following types are of EAPs: **1)** DEs (dielectric elastomers) which are driven by the generated coulomb force, made by sandwiching an elastomer between flexible electrodes. [2, 3], **2)** IPMCs (Ionic polymer-metal composites), which owe their power to the movement of ions and water molecules in the polymer film (combination of an electrolyte film and a thin metal electrode) [4], **3)** CPs (conductive polymer) which use a drive force moving ions by applying a voltage between the conductive polymers [5], **4)** ionic polymer gel Ion polymer gels, which utilize the movement of ions due to chemical changes (e.g., Ph changes)

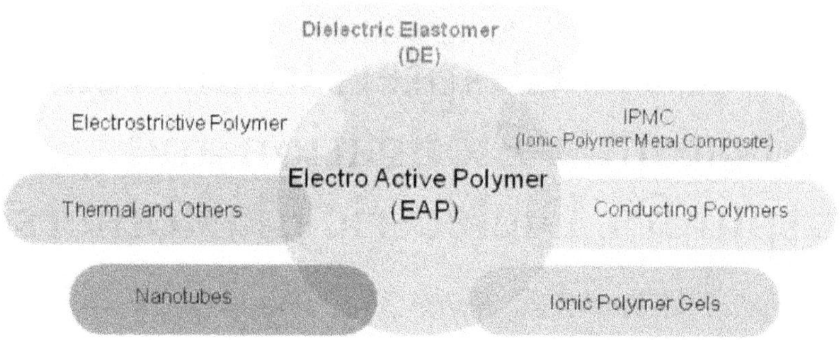

Figure 1.
Typical electroactive polymers (EAP).

within the gel [6], and 5) CNT (carbon nanotube) actuators, which are ideally for nanomachines and do not require ion intercalation [7]. In addition, the piezopolymer utilizes a piezoelectric phenomenon [8], and some are driven by heat, air, light, etc. [9–13].

The most promising candidate from technologies above is the DE [2, 14]. In 1990, Chiba and Pelrine began research and development of dielectric elastomers for the first time in the world. [2], but now, research and development for their practical uses are rapidly progressing all over the world as a technology for practical use [2, 3, 14–34].

Most of the current market demand is for DE power suits, DE motors, DE muscles for robots, and systems that drive them in reverse to generate electricity efficiently. To meet these demands, the elasticity of the elastomer is extremely important. In this paper, we discussed important factors (including cross-linking agents and double bond breaks) through SS (strain stress) curves, viscoelasticity tests, etc. of DE materials. In addition, recent attempts have been made to use new carbon foam materials such as SWCNTs and MWCNTs as electrodes for DEs. These electrodes bring the above-improved elastomers to a higher performance. They will also be discussed in this paper.

2. Background of DEs

The structure of a DE is very simple and consists of a polymer film (elastomer), which is the main material, and two electrodes that sandwich it [2, 3]. When a potential difference is applied between the electrodes, the Coulomb force causes the polymer film to contract in the thickness direction and expand in the plane direction (see **Figure 2**).

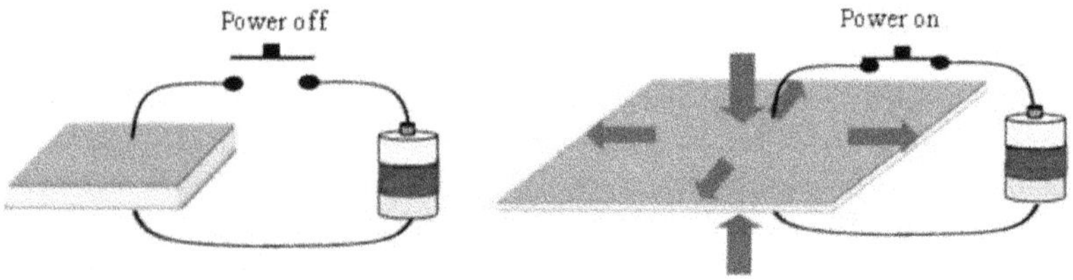

Figure 2.
DE artificial muscle actuator structure and operating principle: (a) The black sheet is the flexible and stretchable electrode, and (b) The yellow part is the elastomer.

At the material level, a DE actuator has a fast response speed (over 100 kHz), with a high strain rate (up to 680%) [15], high pressure (up to 8 MPa), and power density of 1 W/g [16]. A DE actuator having only 0.15 g of DE can lift the weight of 8 kg easily by 1 mm or more with the actuation speed of 88 msec, using Single-wall carbon nanotubes as electrodes [17]. Since the elongation and the output are in inverse proportion to each other, it is possible to suppress the output and increase the elongation. In addition, as mentioned above, power generation is possible by reversing the movement of the DE actuator. Its efficiency is excellent, at over 70% [18].

A mathematical model of the DE actuator can be described as follows: The strain (deformation) observed in the elastomer membrane is mainly caused by the interaction of electrostatic charges between the electrodes [19]. Opposite charges on the two electrodes attract each other, and the same charges repel each other. This phenomenon can be derived by using a simple electrostatic model to derive the effective pressure generated by the electrodes of the elastomer membrane as a function of the applied voltage [19]. Pressure ρ is

$$\rho = \varepsilon_r \varepsilon_o E^2 = \varepsilon_r \varepsilon_o (\mathrm{V}/\mathrm{t})^2 \tag{1}$$

Here, ε_r and ε_o are the permittivity and the relative permittivity (dielectric constant) of the polymer in the free space, respectively, E is the electric field strength, V is the applied voltage, and t is the film thickness. The responsiveness of this polymer is similar to that of conventional electrostrictive polymers, and the pressure is proportional to the square of the electric field strength. For small strains with free boundary conditions, the actuator energy density, e_a, of the material can be written as

$$\mathrm{e_a} = \mathrm{Ps_z} = \mathrm{Ys_z}^2 = (\varepsilon_r \varepsilon_o)^2 (\mathrm{V}/\mathrm{t})^4 / \mathrm{Y} \tag{2}$$

where Y is the modulus of elasticity and s_z is the polymer thickness strain [14]. Conventionally, the elastic energy density e_a = 1/2 $Ys_z{}^2$ is often used (see **Table 1**).

Polymers investigated	**Presure (MPa)**	**Strain (%)**	**Young's modulus (MPa)**	**Breakdown Electric field (V/μm)**	**Dielectric constant (at 1 kHz)**	**Coupling efficiency, *k2* (%)**	**Elastic energy density (Jrcm^3)**
Fluoroelastomer 1	0.11	8	2.5	32	12.7	15	0.0046
Isoprene Natural Rubber 1	0.11	11	0.85	67	2.7	21	0.0059
Silicone 2	0.13	41	0.125	72	2.8	65	0.026
Fluorosilicone 2	0.39	28	0.5	80	6.9	48	0.055
Silicone 3	0.51	32	0.7	144	2.8	54	0.082
Polyurethane 1	1.6	60	17	160	7.0	21	0.087
Silicone 1	1.36	102	1.0	235	2.8	54	0.22
Acrylic 1	7.2	358	2.2	412	4.8	85	3.5

**Average engineering modulus at the maximum strain. Elastic energy density.*

Table 1.
The result of performance measurements of eight polymers (elastomers).

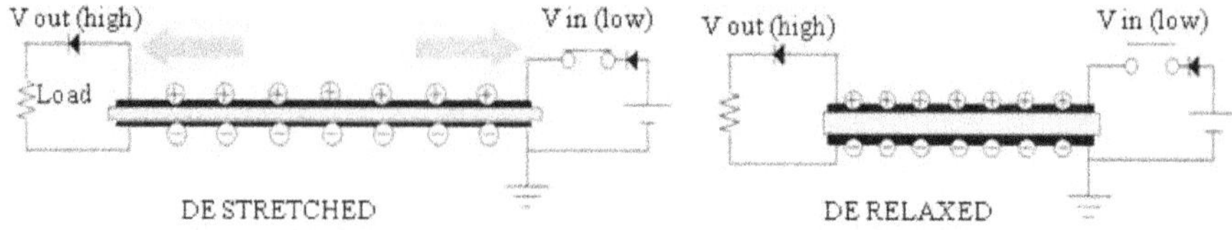

Figure 3.
Operating principle of DE power generation: (a) Thick lines are compliant electrodes, and (b) the yellow line between the thick lines is the dielectric elastomer.

As described above, when the movement of the dielectric elastomer actuator is reversed, the power generation mode is set. This field of power generation research has become more active around the world since [14, 15, 18, 19, 32–56] it was first tested with a DE generator on a buoy [34]. The principle is simple and utilizes the increase in electrostatic energy generated by changing the shape of the dielectric elastomer actuator with an external force (see **Figure 3**). That is, when some mechanical energy is applied to the dielectric elastomer to extend it, the thickness direction becomes thinner and the area expands (Increase in capacitance). At this time, electrostatic energy is generated on the polymer and stored as an electric charge. When the mechanical energy decreases, the elasticity of the dielectric itself increases the thickness in the thickness direction and reduces the area (Reduction of capacitance). At this time, the electric charge is pushed out toward the electrode. Such changes in charge increase the voltage difference, resulting in increased electrostatic energy [19]. The capacitance of the DE film "C" is given as follows:

$$C = \varepsilon_0 \varepsilon A / t = \varepsilon_0 \varepsilon b / t^2 \tag{3}$$

where ε0 is the dielectric permittivity of free space, ε is the dielectric constant of the polymer film, A is the active polymer area, and t and b are the thickness and the volume of the polymer. The second equality in Eq. (3) can be written because the volume of the elastomer is essentially constant, i.e., At = b = constant. The energy output of a DE generator per cycle of stretching and contraction is

$$E = 0.5 C_1 V_b^2 (C_1 / C_2 - 1) \tag{4}$$

where C1 and C2 are the total capacitances of the DE films in the stretched and contracted states, respectively, and V_b is the bias voltage.

3. Materials for DEs

The main parameters that improve the performance of the DE are the withstand voltage of the elastomer film, the dielectric constant (including the improvement of the dielectric constant due to additives), Young's modulus, the type of electrode used, use of a cross-linking agent, and the elastomer structure Improvements (such as the addition of monomers or cutting one of the double bonds).

Table 1 shows the measurement performance of some polymers [3, 20]. This table shows measurements of strain, electric field, modulus of elasticity, and permittivity. The pressure is calculated from Eq. (1) and the elastic energy density is estimated using the strain (measured value) and the pressure calculated from Eq. (2).

Note:

- Silicon 1 was made by mixing two types of silicon polymers.
- Acrylic 1 was modified by us after purchasing acrylic made in the United States.
- The polymers (elastomers) other than the above two types were made in the United States and were used as they were.

As shown in **Table 1**, the DE polymers (elastomers) that can obtain a value with a large strain has a large value of any one of elastic energy density, breakdown electric field, Young's modulus, or permittivity, or some of them are combined thereof. However, increasing these parameters will stiffen the elastomer and will not significantly deform the DE. In other words, the power obtained will not increase unless the elastomer is hardened and deformed (thickness) significantly.

A large deformation is important not only for actuators but also for power generation elements. That is, a large deformation produces more power (see Eq. (4)).

3.1 Elastomer properties obtained from SS curves/dynamic viscoelasticity tests

The SS curve and dynamic viscoelasticity were measured using Silicon 1 and Acrylic 1 [21]. The results are shown in **Figures 4** and **5**. First of all, we would like to point out that the research target is artificial muscles, and it is recommended to consider the tensile speed of the SS curve and the viscoelasticity test from the operating speed required for robots and power assist devices. In **Figure 4**, the SS curve was measured by changing the measurement speed in 4 steps, and the curve changed depending on the tensile speed. Similarly, in **Figure 5**, the curve of dynamic viscoelasticity changed depending on the measurement speed [14].

What is interesting here is that acrylic has higher viscoelasticity, so it depends more on tensile speed than silicon. This indicates that it is important to test it with the response required for the artificial muscle. In other words, until now, researchers have overlooked the importance of viscoelasticity. This meaning is easier to understand by looking at the results of the dynamic viscoelasticity test in **Figure 5**.

As the **Figure 5**, clearly shows, the acrylic is more affected by dynamic viscoelasticity than the silicon. As a result, when the driving voltage is increased and each elastomer is stretched, the silicon DE becomes harder, and the amount of stretching

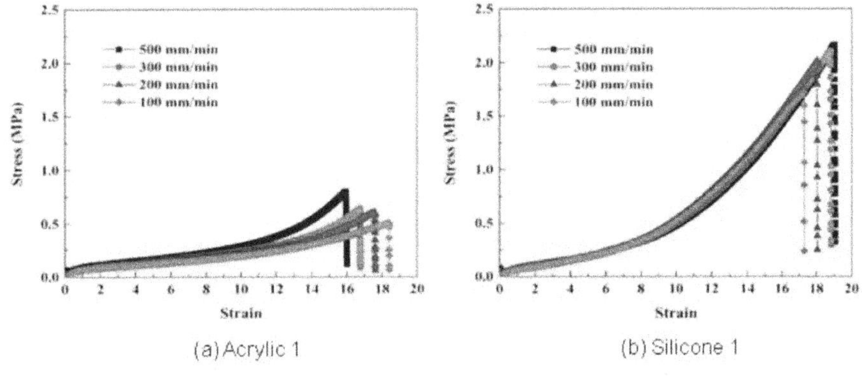

Figure 4.
Relationship of stress-strain for tensile tests: (a) Acrylic 1, (b) Silicone 1.

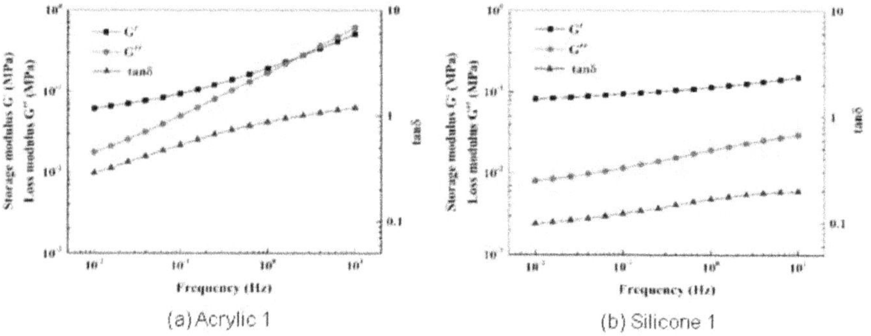

Figure 5.
The frequency dependence of G', G" and tan δ of (a) the acrylic 1 and (b) the silicone 1.

is smaller. Of course, it is also a fact that the difference in the dielectric constant and Young's modulus of both films is the cause (see Eqs. (1) and (2)). Nevertheless, the above behavior can be explained using the SS curves in **Figure 4**. The silicon curve stands up more. This does not mean that silicon has poor performance. It is a proposed that it is better to change the material depending on the application. Silicon has a faster drive speed and a higher rate than acrylic. Therefore, it is advisable to select the type of elastomer depending on where it is used, for example, for applications such as robots or power suits. Illustrating this point, in human muscles, there are slow muscles and fast muscles, each of which has an important mission. Moreover, Silicon can be used from a relatively high temperature to a considerably low temperature. Compared to acrylic, it could have a considerable advantage in devices that are used at higher or lower temperatures [21].

In terms of artificial muscles, e.g., for human uses, it seems that a flat hill-like shape with a gentle rise, like acrylic, is preferable.

3.2 Attempts to increase the dielectric constant of elastomers

As an attempt to increase the dielectric constant of the elastomer, there is a method of adding a monomer to change the structure or adding a substance having a very high dielectric constant such as Barium titanate. However, in general, these methodss make the elastomer harder and less stretchable. Examples of adding Barium titanate to our synthetic acrylic are shown below. Here, the acrylic we have synthesized is called a base acrylic.

As a method for dispersing barium titanate, a predetermined amount of barium titanate was added to the polymer-containing liquid and crushed with a homogenizer. As a result of particle size measurement by SEM, the median diameter was about 450 nm [57]. It was also confirmed by using SEM that the barium titanate was uniformly mixed.

Elastomer sheets (thickness: 400 μm) were prepared by a) adding 1 wt% of Barium titanate to the base acrylic, and b) adding a 2 wt% of Barium titanate to the acrylic. The SS curves of those elastomers were measured as shown **Figure 6**.
The acrylics, which are the base for the Barium titanate were slightly crosslinked. The permittivity of the acrylic was measured using the parallel plate capacitance method. The withstand voltage was measured using a general dielectric breakdown tester. The relationship between the withstand voltage and the capacitance of these films is shown in **Figure 7**.

From **Figures 6** and 7, as we initially expected, the withstand voltage and the amount of capacitance of the film containing a large amount of Barium titanate increased, and the film became harder and less stretchable by that amount. Circular

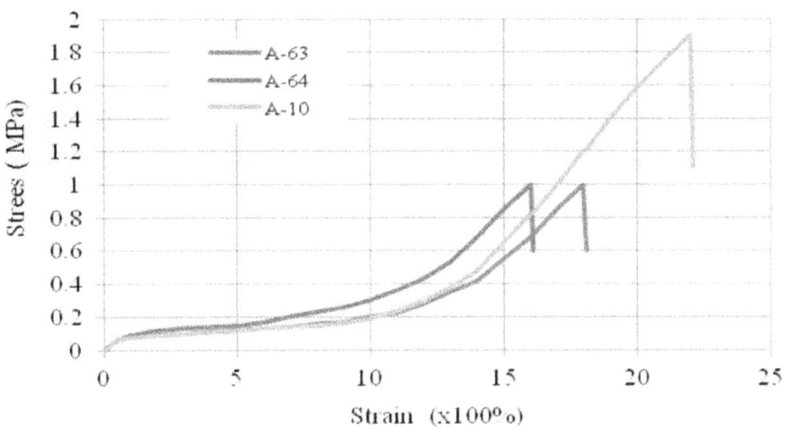

Figure 6.
The SS curve of the elastomer sheet with a small amount of a) 1wt% and b) 2wt% of Barium titanate added to the base acrylics.

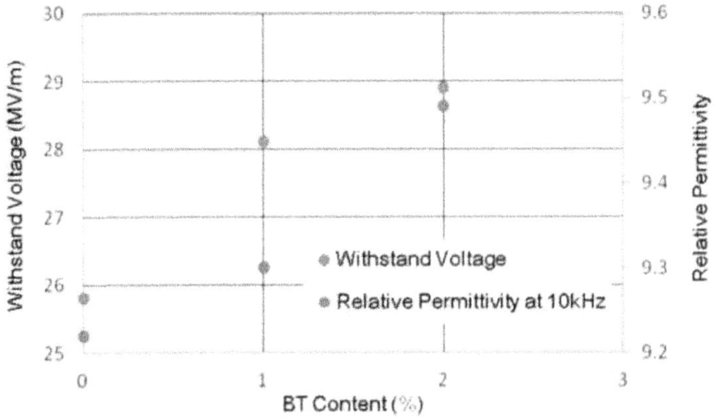

Figure 7.
Relationship between the breakdown of the electric field and the capacitances of these elastomer films.

actuators were produced using either the base acrylic film without adding the barium titanate, or the films containing 1 wt% or the barium titanate and 2 wt%, and the elongations of each were compared. As a result, the actuator using the base film showed the largest elongation. In fact, the film that was hardened by adding the barium titanate was superior in increasing the withstand voltage. From those results, it was found that even if a substance with a high dielectric constant was added, it did not give a significantly good effect. The dynamic viscoelasticity of the base acrylic +2% of Barium titanate is shown in **Figure 8** (**Figure 8** is in Section 3.3).

3.3 Adjustment of cross-linking agents/reduction of double bonds

Figure 8 shows the SS curve when the amount of cross-linking agent added to the above base acrylic is changed. Assuming that the amount of the cross-linking agent added to the base acrylic (blue) is 1, red, green, black, and orange are added at rations of 2:1, 1.5:1, 0.8:1, and 0.5:1, respectively. Not surprisingly, the less cross-linking agent we add, the better the elongation. Due to that reduced strength, circular actuators need to be manufactured and evaluated to determine how appropriate they are.

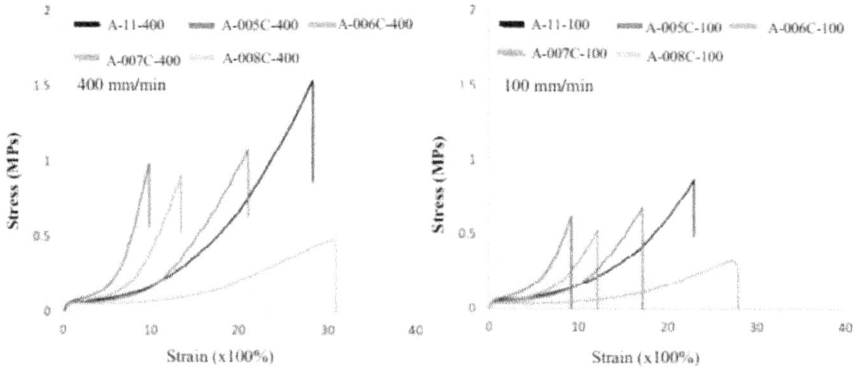

Figure 8.
The SS curve when the amount of cross-linking agent added to the above base acrylic changed. The case where the amount of the cross-linking agent added was changed and the case where the amount of the double bond was reduced by using HNBR (Hydrogenated acrylonitrile butadiene rubber).

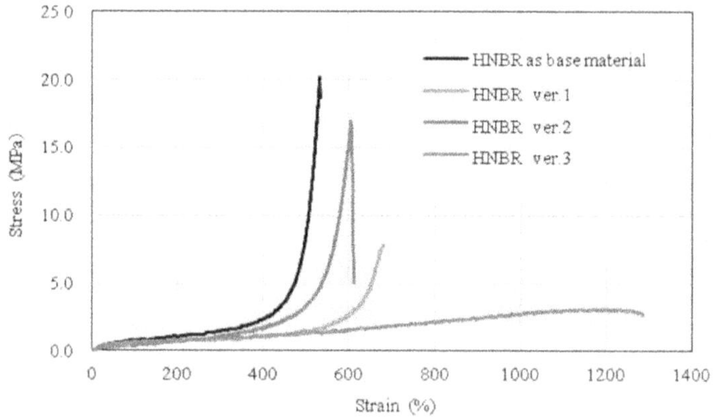

Figure 9.
The SS curve when the amount of cross-linking agent added to the above base acrylic changed: The case where the amount of the cross-linking agent added was changed and the case where the amount of the double bond was reduced by using HNBR (Hydrogenated acrylonitrile butadiene rubber).

In this experiment as well, the tensile speeds were set to 100 mm / min and 400 mm / min, but as mentioned above, such an evaluation is important for artificial muscles. That is, the test should be performed according to the actual running speed of the muscle. In this case, 100 mm / min / sec is clearly affected by viscoelasticity. In other words, even if the elongation increases, the stress does not increase so much (the inclination angle is gentle), and as a result, it could become easy to deform as the DE.

Next, attempts were made to not only change the amount of additive added, but also to reduce the amount of double bonds. **Table 2** shows the conditions of the case where the amount of the cross-linking agent added was changed and the case where the amount of the double bond was reduced by using HNBR (Hydrogenated acrylonitrile butadiene rubber). HNBR is a hard material with a dielectric constant of 15, but as shown in **Table 2**, when the ratio of double bonds is reduced, the slope of the SS curve becomes gentle (see **Figure 9**). HNBR Ver.3 has a dramatically reduced slope because the cross-linking agent has also been reduced. The capacitance was also 11.

	Crosslinker (%)	Double bond (%)
HNBR sheet as Base Material	8	10
HNBR ver.1	8	5
HNBR ver. 2	8	1
HNBR ver. 3	2	1

Table 2.
The case where the amount of the cross-linking agent added was changed and the case where the amount of the double bond was reduced by using HNBR (hydrogenated acrylonitrile butadiene rubber).

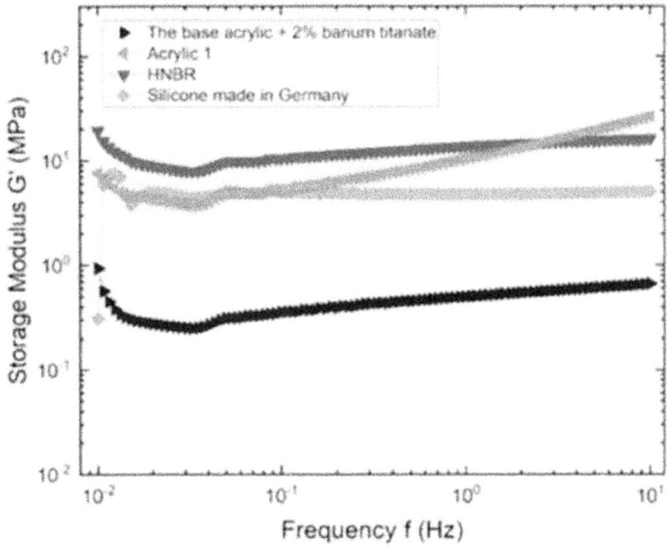

Figure 10.
The SS curves of HNBER (base material), HNBER ver.1, HNBR ver.2 and HNBER ver.3. Note: The silicon for this test was used the silicon German-made. This is because the silicon 1 was tested long time ago there is no remaining stock.

With these membranes, it is necessary to make a circular actuator and measure the elongation, but unfortunately it has not been done yet. Perhaps HNBR ver.3 is a little too soft and it could be difficult to make a DE. Or, because it is soft, the Coulomb force might be dispersed and it might not be possible to drive it well. Further studies are desired on the proportion of double bonds and the amount of cross-linking agent.

The dynamic viscoelasticity of HNBR was also measured (see **Figure 10**). Silicon and acrylic are also shown in this figure for comparison.

Figure 10 shows the frequency dependence of the storage modulus of five kinds of materials. From 0.032 Hz, it can be seen that the storage elastic modulus of the acrylic 1 (see **Table 1**) increases. It can be seen that the storage elastic modulus of HNBR ver.3 gradually increases, but the storage elastic modulus of the silicon (made in Germany) basically does not change as the frequency increases. Again, silicon could be a bit difficult to get the most out of as an artificial muscle. One of the reasons might be that the structure of silicon is generally a chain structure. Of course, silicon has excellent temperature characteristics and DE responsiveness, and can be driven efficiently.
As for our recommendation, it is a good idea to use both fast (silicon) and slow (acrylic) muscles well, like human muscles. Since HNBR is rubber, it is resistant to humidity and can withstand temperature changes. In addition, the results of dynamic

viscoelastic research show that it is somewhat suitable for driving a DE. In particular, it seems that the amount of cross-linking agent added should be selected appropriately. We believe it is particularly suitable for ocean power generation. Since ocean power generation is exposed to a harsh natural environment, it is desirable to use a material that is tough and suitable for DEs. Another point that greatly contributes to power generation efficiency is that there are many changes in thickness (see Eq. (3)), and in that respect, acrylic is most suitable, but acrylic is not very suitable for harsh natural environments. We hope that moisture resistant acrylics will be developed. The film with 2% barium titanate added to the base acrylic is considerably harder than the other films as mentioned above, but the withstand voltage of this film is high (see **Figure 6**) and the elastic modulus is also increased. Therefore, if it could withstand a higher voltage, it might be used as a high-power DE in the future.

3.4 Pre-strain

Pre-strain will increase the performance of the DE. This is because when repeated tests were performed to know the SS curve of the elastomer sample, the film stretched and it could not return to its original length, so there was no choice but to stretch the film a little in advance and evaluate it [14, 20]. After that, if more pre-strain is applied, it will be advantageous because the strain is applied in advance compared to the case where it is not stretched, and the performance will be further improved [34]. In order to utilize the pre-distortion, it is advantageous to use a material having a gentle SS curve, such as acrylic (see **Figure 4**). As described above, as the degree of pulling increases, the film becomes harder and harder to stretch. However, since the curve of acrylic is gentle, it is harder to harden than silicon. In the dynamic viscoelasticity test, acrylic is more frequency dependent and has more storage modulus than silicon (see **Figures 5** and **8**). This means that even if the film becomes hard, it can function as a DE because the modulus increases. Again, this frequency dependence is also important for use as an artificial muscle.

3.5 Adopted CNTs as electrodes

Recently, attempts have been made to use new carbon foam materials such as SWCNTs and MWCNTs as electrodes for DEs. These electrodes bring the elastomers to a positively improved higher performance. **Table 3** shows how much weight can be lifted with a stroke of 5 mm due to the difference in electrodes. The elastomer used is acrylic 1, and its weight is 0.1 g [10]. Diaphragm actuators with a diameter of 8 cm were manufactured and those elongations were measured.

Since these electrodes are not optimized, it seems possible to lift heavier weights while having sufficient elongation in the very near future. In addition, these are single layers of DEs and are very light, so it is possible to have multiple layers of DEs, which is close enough to the range applicable to robots and power suits.

On the other hand, these electrodes are also promising as power generation elements. A power generation experiment was conducted using a drape type DE having a height of 120 mm and a diameter of 260 mm. The amount of power generation when the DE was pulled by about 60 mm is summarized in **Table 4** [15]. The drape weighs 4.6 g and uses acrylic 1. Carbon grease, Carbon black, MWCNTs (multi-walled carbon nanotubes), and SWCNTs (single-walled carbon nanotubes) were used as electrode materials.

Using MWCNTs or SWCNTs makes it possible to obtain more power, as shown in **Table 4**.

Electrode type	Weight that can be lifted with a stroke of 5 mm
Carbon grease	6.5 N
Carbon Black	10 N
Multi-walled carbon nanotubes (MWCNT)	16 N
Single-walled carbon nanotubes (SWCNT)	22 N

Table 3.
Types of electrodes and weight that can be lifted.

Type of electrode	Power obtained (mJ)
Carbon grease	179
carbon black*	274
multi-walled carbon nanotube	445
single-walled carbon nanotube**	630

**Carbon grease, Carbon black and MWCNT are manufactured in companies in United States*
***SWCNT (ZEONANO®-SG101) by Zeon Corp in Japan.*

Table 4.
Differences in power obtained when changing the electrode materials.

This is because the conductivities of MWCNTs and SWCNTs are much higher than that of carbon black or Carbon grease.

In this way, the highly conductive material significantly increases the elongation of the DE actuator, resulting in greater elongation and also increasing the amount of power generated by the DE element.

4. Conclusion

From the above experimental results and those discussions, the following was found:

- The elongation of the DE is greatly influenced by the elastomer material. If the material is too hard or too soft, it may not produce the desired result. DE performance needs to be adjusted according to product requirements. The same applies to the output of the DE.

- There are several methods for changing the properties of the elastomer, but two easy methods are to adjust the amount of the cross-linking agent added and/or the percentage of double bonds. Mixing additives with a high dielectric constant seems to be less effective.

- In order to know the properties of the polymer materials for the DE, it is desirable to measure the SS curve and/or dynamic viscoelasticity. In those cases, it is recommended to measure at a slower speed for artificial muscle applications.

- By using highly conductive materials, it is possible to improve the performance of the DE actuator and the performance of the DE power generator.

Acknowledgements

We thank Zeon Corporation for supplying SWCNT ZEONANO®-SG101 and HNBR.

Author details

Seiki Chiba[1*], Mikio Waki[2], Shijie Zhu[3], Tonghuan Qu[3] and Kazuhiro Ohyama[3]

1 Chiba Science Institute, Yagumo, Meguro ward, Tokyo, Japan

2 Wits Inc., Oshiage, Sakura, Tochigi, Japan

3 Fukuoka Institute of Technology, Wajirohigashi, Higashi-ward, Fukuoka, Japan

*Address all correspondence to: epam@hyperdrive-web.com

References

[1] Roentgen, W. About the Change in Shape and Volume of Dielectrics Cased by Electricity, Ann. Phys. Chem., Vol.11, pp.771-788, 1880

[2] Pelrine, R.; Chiba, S. Review of Artificial Muscle Approaches, Proc. of Third Intl Symposium on Micromachine and Human Science, Japan, 1992.

[3] Pelrine, R.; Kornbluh, R.; Chiba, S. et al. High-field defomation of elasomeric dielectrics for actuators, Proc. 6th SPIE Symposium on Smart Structure and Materials, Vol. 3669, 1999, pp-149-161.

[4] Oguro, K. et al. Polymer electrolyte actuator with gold electrodes, Proc. SPIE, Vol. 3669, 1999,

[5] Otero, T.; Sansiñena, J. Soft and wet conducting polymers for artificial muscles, Advanced Materials 10 (6), 1998, pp. 491-494.

[6] Osada, Y. et al. A polymer gel with electrically driven motility, Nature, Vol. 355, 1992, pp. 242-244.

[7] Baughman R. et al. Carbon Nanotube Actuators, Science, Vol. 284, (1999) pp. 1340-1344.

[8] Gross, B. Experiments on Electrets, Phys. Rev. 66, 26, 1944

[9] Chou, P.; Hannaford. B. Static and dynamic characteristics of mckibben pneumatic artificial muscles," Proceedings of the IEEE International Conference on Robotics and Automation, SanDiego, CA, May 8-13, 1994, p.281-286.

[10] Smots, G. New developments in photochromic polymers, J. Polymers Science, Polymers Chemistry, 1995, Vol. 13, p.2223.

[11] Tobushi, H.; Hayashi, S.; Kojima, S.; Mechanical Properties of Shape Memory Polymer of Polyurethane Series, in JSME International J., 1992, Series 1, Vol. 35, No. 3.

[12] Bar-Cohen Y. et al. Electroactive Olymer (EAP) Actuators as Artificial Muscles – Reality Potential and Challenges",1st edition, ed. Y. Bar-Cohen, SPIE Press (2001-03)

[13] Ratna, B. Liquid crystalline elastomers as artificial muscles: role of side chain-backbone coupling, Proc. SPIE, 2001.

[14] Waki, M.; Chiba, S. Application of Dielectric Elastomer Transducer, Chapters 33, Materials, Composition and Applied Technology of Soft Actuators, S & T Publishing, 2016, ISBN978-907002-61-9 C3058.

[15] Chiba, S.; Waki, M.; Jiang, C.; Takeshita, M.; Uejima, M.; Arakawa, K.; Ohyama, K. The Possibility of a High-Efficiency Wave Power Generation System using Dielectric Elastomers, Energies, 2021, 14, 3414. https://doi.org/10.3390/en14123414

[16] Chiba, S.; Stanford, S.; Pelrine, R.; Kornbluh, R.; Prahlad, H.; "Electroactive Polymer Artificial Muscle", JRSJ, Vol. 24, No.4, pp 38-42, 2006.

[17] Chiba S et al. Challenge of creating high performance dielectric elastomers, Proc. of SPIE2021 (Smart Structures and Materials Symposium and its 23rd Electroactive Polymer Actuators and Devices (EAPAD) Conference); 2021:1157-62.

[18] Chiba, S.; Waki, M.; Wada, T.; Hirakawa, Y.; Masuda, K.; Ikoma, T. Consistent ocean wave energy harvesting using electroactive polymer (dielectric elastomer) artificial muscle generators, Applied Energy, Elsevier, Volume 104, April 2013, Pages 497-502, ISSN 0306-2619.

[19] Chiba, S. Dielectric Elastomers, Chapter 14, Soft actuators, 2nd Edition, Springer Nature, 2019; https://doi.org/10.1007/978-981-13-6850-9

[20] Chiba, S.; Waki, M.; Application to dielectric elastomer materials, power assist products, artificial muscle drive system. In Next-Generation Polymer/Polymer Development, New Application Development and Future Prospects; Technical Information Association: Tokyo, Japan, 2019; Section 3, Chapter 4. ISBN-10: 4861047382 ISBN-13: 978-4861047381

[21] Li, W.; Zhu, S.; Ohyama, K.; Chiba, S.; Waki, M. Mechanical properties and viscoelasticity of dielectric elastomers. In Proceedings of the Materials and Mechanics Conference, Japan Society of Mechanical Engineers, Sapporo, Japan, October 7-9, 2017; No. 17-7, GS0302.

[22] Hu, P.; Madsen, J.; Skov. A. Super-stretchable silicone elastomer applied in low voltage actuators," Proc. SPIE 11587, Electroactive Polymer Actuators and Devices (EAPAD) XXIII, 1158715 (22 March 2021); doi: 10.1117/12.2581476.

[23] Kumamoto, H.; Hayashi, T.; Yonehara, Y.; T. Okui.; Nakamura, T. Development of Development of a locomotion robot using deformable dielectric elastomer actuator without pre-stretch, Proc. SPIE 11375, Electroactive Polymer Actuators and Devices (EAPAD) XXII, 1137509 (22 April 2020); doi: 10.1117/12.2558422

[24] Carmel, M. Enhancing the permittivity of dielectric elastomers with liquid metal, Proc. of SPIE 11375-15, Electroactive Polymer Actuators and Devices (EAPAD) XXII, 22-26 April 2020.

[25] Albuquerque, F.; Shea, H. Effect of humidity, temperature, and elastomer material on the lifetime of silicone-based dielectric elastomer actuators under a constant DC electric field, Proc. of SPIE 11375-42, Electroactive Polymer Actuators and Devices (EAPAD) XXII, 22-26 April 2020.

[26] Jayatissa, S.; Shim, V.; Anderson, I.; Rosset, S. Optimization of prestretch and actuation stretch of a DEA-based cell stretcher, Proc. of SPIE 11375-50, Electroactive Polymer Actuators and Devices (EAPAD) XXII, 22-26 April 2020.

[27] Hu, P.; Huang, Q.; Madsen, J.; Skov, A. Soft silicone elastomers with no chemical cross-linking and unprecedented softness and stability," Proc. SPIE 11375, Electroactive Polymer Actuators and Devices (EAPAD) XXII, 1137517 (24 April 2020); doi: 10.1117/12.2557003

[28] Kunze, J.; Prechtl, J.; Bruch, D.; Nalbach, S.: Motzki, P.; Mechatronik, Z.; Seelecke, S. Design and fabrication of silicone-based dielectric elastomer rolled actuators for soft robotic, applications, Proc. of SPIE 11375-80, Electroactive Polymer Actuators and Devices (EAPAD) XXII, 22-26 April 2020.

[29] Ichige, D.; Matsuno, K.; Baba, K.; Sago, G.; Takeuchi, H. A method to fabricate monolithic dielectric elastomer actuators, Proc. of SPIE 11375-81, Electroactive Polymer Actuators and Devices (EAPAD) XXII, 22-26 April 2020

[30] Pei Q., Dielectric Elastomers past, present and potential future, Proc. of SPIE, 10594-4, 2018.

[31] Kornbluh, R.; Pelrine, R.; Prahlad, H.; Wong-Foy, A.; McCoy, B.; Kim, S.; Eckerle, J.; Low, T. Promises and Challenge of dielectric elastomer energy harvesting, Proc. of SPIE Vol. 7976, 79605, Electroactive Polymer Actuators and Devices (EAPAD) 2011, doi:10.1117/12.882367.

[32] Chiba, S.; Waki, M. Recent Advances in Wireless Communications

and Networks, Wireless Communication Systems, Chapter 20, pp. 435-454, InTech, 2011.

[33] Chiba S. et al. "Elastomer Transducers", Advances in Science and Technology, Trans Tech Publication, Switzerland, Vol. 97, pp 61-74, 2016, ISSN: 1662-0356, Doi: 10.4028/www scienctific.net/AST.97.61

[34] Chiba, S.; Pelrine, R.; Kornbluh, R.; Prahlad, H.; Stanford, S.; Eckerle, J. New Opportunities in Electric Power Generation Using Electroactive Polymers (EPAM), Journal of the Japan Institute of Energy, Vol. 86, No.9, pp 743-737, 2007

[35] Koh S. J.; Zhao, X.; Suo, Z. Maximal Energy That Can Be Converted by a Dielectric Elastomer Generator. Appl. Phys. Lett. 2009, 94 ,262902-26903.

[36] Yurchenko, D.; Lai, Z.; Thomson, G.; Val, D.; Bobryk, R. Parametric study of a novel vibro-impact energy harvesting system with dielectric elastomer. Appl. Energy 2017, 208, 456-470, doi:10.1016/ j.apenergy.2017.10.006.

[37] Jean-Mistral, C.; Basrour, S.; Chaillout, J.-J. Dielectric polymer: Scavenging energy from human motion. Electroact. Polym. Actuators Devices EAPAD 2008 2008, 6927, 692716.

[38] Chiba, S.; Waki. M.; Fujita. K.; Song. Z.; Ohyama. K.; Zhu. S.; Recent Progress on Soft Transducers for Sensor Networks. In Technologies and Eco-Innovation toward Sustainability II; Hu, A.H., Ed.; Springer Nature: London, UK, 2019; doi. org/10.1007/981-13-1196-3_23.

[39] Jiang, C.; Chiba, S.; Waki, M.; Fujita, K.; Moctar, O. An Investigation of Novel Wave Energy Generator Using Dielectric Elastomers. In Proceedings of the ASME 2020 39th International Conference on Ocean, Offshore and Arctic Engineers, Virtual, Online, 3-7 August 2020, DOI.org/ 10.1115/ OMAE2020-18106.

[40] Chiba, S.; Waki, M.; Innovative power generator using dielectric elastomers (creating the foundations of an environmentally sustainable society), Sustainable chemistry and Pharmacy, Elsevier: Amsterdam, The Netherlands, 15, 2020, 100205

[41] Carpi, F.; Anderson, I.; Bauer, B.; Frediani Gallone, G.; Gei, M.; Graaf, C. Standards for dielectric transducers. Smart Mater. Struct. 2015, 24, 105025.

[42] Jean-Mistral, C.; Basrour, S.; Chaillout, J.-J. Comparison of electroactive polymers for energy scavenging applications. Smart Mater. Struct. 2010, 19, 085012, doi:10.1088/0964-1726/19/8/085012.

[43] Zhong, X. Dielectric Elastomer Generators for Wind Energy Harvesting. Ph.D. Thesis, University of California Los Angeles, Los Angeles, CA, USA, 2010.

[44] Huang, J.; Shian, S.; Suo, Z.; Clarke, D. Maximizing the Energy Density of Dielectric Elastomer Generators Using Equi-Biaxial Loading. Adv. Funct. Mater. 2013, 23, 5056-5061.

[45] Brouchu, P.A.; Li, H.; Niu, X.; Pei, Q. Factors Influencing the Performance of Dielectric Elastomer Energy Harvesters. In Proceedings of the SPIE, 2011.

[46] Moretti, G.; Fontana, M.; Vertechy, R. Parallelogram-shaped Dielectric Elastomer Generators: Analytical model and Experimental Validation. J. Intell. Mater. Syst. Struct. 2015, 26, 740-751.

[47] Kovacs, G.M. Manufacturing polymer transducers: Opportunities and challenges. Proc. SPIE 2018, 2018, 10594-10597.

[48] Pelrine, R.; Kornbluh, R.D.; Pei, Q. Dielectric elastomers: Past, present, and potential future. Electroact. Polym. Actuators Devices EAPAD XX 2018, 10594, 1059406, doi:10.1117/12.2302815.

[49] McKay, T.; O'Brien, B.; Calius, E.; Anderson, I. Soft Generators Using Dielectric Elastomers. Appl. Phys. Lett. 2011, 98, 1-3.

[50] Anderson, I.; Gisby, T.; McKay O'Brienl, B.; Calius, E. Multi-functional Dielectric Elas T.tomer Artificial Muscles for Soft and Smart Machines. J. Appl. Phys. 2012, 112, 041101.

[51] Kessel, V.; Wattez, R.; Bauer, P.; Analyses and Comparison of an Energy Harvesting System for Dielectric Elastomer Generators Using a Passive Harvesting Concept: The Voltage-clamped Multi- phase System. In Proceedings of the SPIE Smart Structures and Materials+ Nondestructive Evaluation and Health Monitoring, International Society for Optics and Photonics, 2015; 943006.

[52] Chiba, S.; Waki, M.; Fujita, K.; Masuda, K.; Ikoma. T. Simple and Robust Direct Drive Water Power Generation System Using Dielectric Elastomers. Journal of Material Science and Engineering B7 2017 (1-2): 39-47. DOI: 10.17265/2161-6213/2017.1-2.005.

[53] Chiba, S.; Hasegawa. K; Waki, M.; Fujita, K.; Ohyama, K.; Shijie, Z. Innovative Elastomer Transducer Driven by Karman Vortices in Water Flow, Journal of Material Science and Engineering A7 (5-6), 2017: 121-135. DOI: 10.17265/2161-6213/2017.5-6.002

[54] Waki, M.; Chiba, S.; Song, Z.; Ohyama, K.; Shijie, Z. Experimental Investigation on the Power Generation Performance of Dielectric Elastomer Water Power Generation mounted on a Square Type Floating Body, Journal of Material Science and Engineering B7 (9-10), 2017, 179-186, DOI:10.17265/2161-6221/2017.9-10.001

[55] Chiba, S.; Waki, M.; Masuda, K.; Ikoma, T.; Osawa, H.; Suwa, Y. Innovative Power Generation System for Harvesting Wave Energy, Design for Innovative Value towards a Sustainable Society. Netherlands: Springer, 1002-7, ISBN 978-94-007-3010-6.

[56] Chiba, S.; Waki, M.; Jiang, C.; Fujita, K. Possibilities for a Novel Wave Power Generator Using Dielectric Elastomers, In Proceedings of the ASME 2020 39th International Conference on Ocean, Offshore and Arctic Engineers, Virtual, Online, 3-7 August 2020, DOI. org/10.1115/OMAE2020-18464.

[57] Noro, J.; Kato, J. Analysis of Surface area, Pore Size distribution, and Particle Size Distribution, Analysis, 2009, 7. 349-355.

Chapter 8

Compression and Recovery Functional Application for the Sportswear Fabric

Ramratan Guru, Rajeev Kumar Varshney and Rohit Kumar

Abstract

A sportswear fabric should have good stretch and recovery behaviour. This study facilitates an effective design and development of high-stretch sportswear using different knitted structure. Nine types of knitted fabrics were produced by varying the type of fibre and type of structure. An experiment work is done to study the fabric size, stretch and elastic recovery properties. The statistical analysis showed that type of fibre and type of knitted structure significantly influence the fabric stretch. Plain structure fabric showed higher stretch value than rib and interlock-knitted fabric. The high stitch density caused by reduce stretch value in the course- and wale-wise due to yarn floating rather than overlapping influenced the weight and thickness of knitted fabrics. The elastic recovery analysis indicated that the recovery value of plain-knitted structure with polyester-spandex blend is higher among studied fabrics. However, the recovery value decreased over time in comparison with stretch value.

Keywords: sportswear fabric, stretch and recovery performance, polyester, micro-polyester, polyester-spandex

1. Introduction

The stretchable knitted structures play an important role in body comfort and fit. The knitted structures allow wearer the freedom of movement with least resistance due to their stretchability and elasticity [1, 2]. Regular physical activity is important to maintain consistency in human health. To achieve comfort and functional support during various activities such as walking, stretching, jogging, athletes and sports persons use sports clothing [3, 4]. Stretch properties represent a significant mechanical property of clothing material that influences clothing pressure. Stretch properties are measured as the percentage of fabric stretch and fabric growth, and recovery [5–7].

Basically, two types of category are normally available in sportswear stretch cloths: First is comfort stretch range of about 20–30% and other power stretch cloth range approximately 30–50%. The basic designs are used for high-active sportswear garments in elasticity and compression cloths.

The power stretch cloths need to have more extensibility and quicker recovery performance [8–12].

The high-compression clothes are more utilized medical compression garments and sportswear cloths sectors. Study on the evaluation of elastic recovery of cotton-knitted fabrics was conducted [13–16]. It is found that length of these cloths is different to a more extent value. As per information, fabric elongation is different from single-structure knit fabric in lengthwise ranging 3–6%, with double cloth knit having lengthwise 3–50%. Elongation in single-structure knit fabric is of widthwise 3–180% and double cloth knit elongation in widthwise of 6–155%. It is elastic recovery different from single-structure knit lengthwise to other structures. According to author, single-knit lengthwise elastic recovery is of range 100–56%, double cloth knit lengthwise elastic recovery is of range 100–57%, and single-structure knit elastic recovery widthwise is found 100–56%, double cloth knit widthwise elasticity recovery is of range 100–30% that is basically found in knit cloths [13–17].

Plain knit had more elongation and growth as compared with double knit. The growth after 30 seconds or relaxation was observed to be 36% and plain knit stretched more under load and after the load was released that exhibited more growth than the double knit. The stretch of knit fabrics is affected more by the cover factor than by the yarn diameter, loop length, loop density or the shape of the loop [18]. Spandex is widely used in sportswear for its superior stretch and recovery properties. Dynamic elastic repossession can assess the immediate apparel response due to body movement; the elastic bare-plaited fabric is found to have higher dynamic elastic recovery than cloth knitted from lycra core spun. The basic phenomena are essential use in stretch and recovery of the cloth to pressure generated by compression apparel. It is found that knitted fabric in normal stretch and recovery performance as compared to compression sportswear garment. Therefore, Lycra is used in knit cloths in blend with other fibres for proper utilization of stretchability and elasticity recovery properties in sportswear garments [19, 20].

The objective of this study was to investigate the effect of the stretch, growth fabric and recovery properties of polyester-spandex-blended, micro-polyester and 100% polyester-knitted fabric. These works could facilitate the design and development of sportswear with the required stretch and recovery properties.

2. Materials and methods

2.1 Materials used

In this study, three different filament yarns—polyester, micro-polyester, blend of polyester-spandex and non-circular cross section were used to prepare samples. The knitted structures—single jersey, interlock and rib fabrics were produced on weft-circular knitting machine (**Table 1**).

2.2 Testing methods

The knitted fabric was conditioned in standard atmospheric condition of 65+/−2% RH and 27+/−2°C temperature and the samples are in condition for 24 hours before testing. The stretch and recovery property tester was using the ASTM D 2594-2004 (2008) standard.

2.2.1 Statistical analysis

One-way ANOVA (Minitab 17 statistics software) tests were used to determine the significant difference between the stretch and elastic recovery properties of fabrics. In order to infer whether the parameters were significant or not, p values were examined. If the 'p' value of a parameter is greater than 0.05 ($p > 0.05$), the parameter was not significant and should not be investigated.

Sample	Fabric structure	Wpcm	Cpcm	Stitch density (loop/cm^2)	Areal density (g/m^2)	Thickness (mm)	Loop length (cm)	Tightness factor (Tex$^{1/2}$/cm)
PET/spandex	Plain	14.17	20.83	295.16	115	0.57	0.39	8.53
	Rib	16.12	24.13	388.97	155	0.65	0.32	10.74
	Interlock	19.31	26.18	527.54	178	0.81	0.26	12.80
Micro-PET	Plain	17.28	15.43	266.63	105	0.50	0.39	8.53
	Rib	16.12	20.14	324.65	144	0.58	0.31	10.74
	Interlock	22.67	17.12	388.11	161	0.68	0.26	12.80
100% PET	Plain	17.12	13.21	226.15	98	0.41	0.39	8.53
	Rib	15.32	20.82	318.91	136	0.52	0.32	10.74
	Interlock	22.15	16.83	372.78	155	0.62	0.26	12.80

**Wpcm—wale per centimetre, Cpcm—course per centimetre, * PET—polyester, * all three fabrics are made from 100 deniers polyester tex 11.11.*

Table 1.
Geometrical properties of knitted fabrics.

2.2.2 Fabric particulars

The fabric details measured were as follows: wales per inch (wpi), course per inch (cpi), linear density of yarn (denier), fabric mass per unit area (g/m^2) and fabric thickness (mm). The wpi and cpi were measured according to the ASTM D-3887 Standard. Yarn linear density and fabric mass per unit area were measured according to ASTM D 1059-01 and ASTM D 3776 M-09a standard respectively by using an electronic weighing balance. Thickness testing was carried out as per BS EN ISO 9073-2 using the electronic thickness tester at 0.25-KPa pressure. For each sample, 30 readings were taken to get the result at 95% confidence level.

2.2.3 Stretch and recovery property tester

ASTM D 2594-2004 (2008) standard test method for stretch properties of knitted fabric was applied under the form-fitting standards as sample size 18.3 cm in wale direction and 21.5 cm in course direction positions and 4.54 kgf weight apply in the Fabric specimen of the lower bench marks were calculated after 60 s and 1 h. Stretch percentage points to test the stretch and elastic recovery of the experimental samples are as shown in **Figures 1** and **2.**

Stretch, growth and recovery percentages were calculated by Eqs. (1)–(3) given below:

$$\text{Fabric stretch\%} = \frac{\text{B} - A}{\text{A}} X100 \tag{1}$$

$$\text{Fabric growth\%} = \frac{\text{C} - A}{\text{A}} X100 \tag{2}$$

$$\text{Fabric recovery\%} = \frac{\text{B} - C}{\text{B} - \text{A}} X100 \tag{3}$$

where A: the distance marked between the upper and bottom parts of the fabric; B: the distance between the marked points after holding the sample for 5 min with 4.54 Kgf load; C: the distance between the marked points after 5-min relaxation.

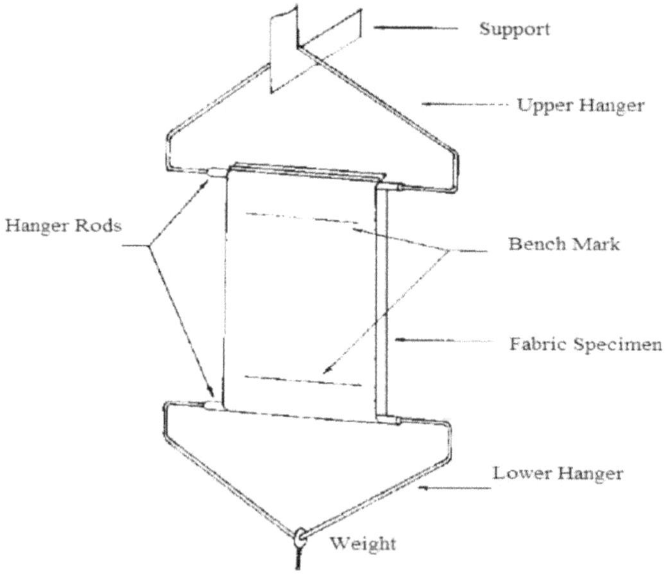

Figure 1.
Stretch and recovery setup assembly.

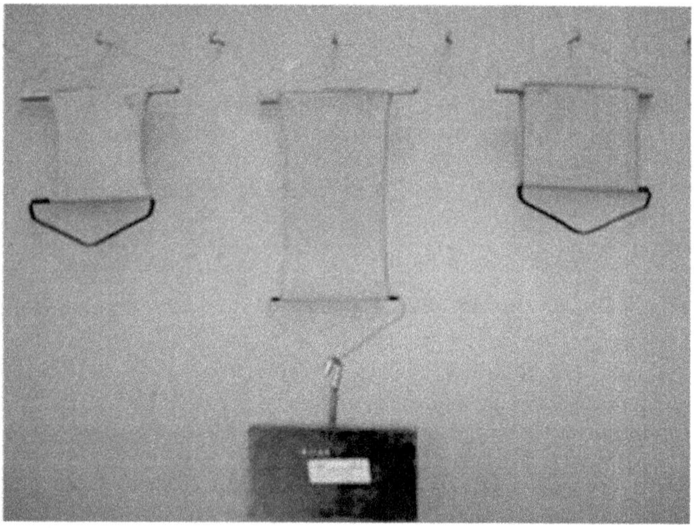

Figure 2.
Fabric stretch equipment.

3. Results and discussion

Figures 3 and **4** and show knit specimen changes in weight and thickness. The heavier weight of knit specimen was, the thicker its thickness was in descending order 'interlock structure polyester-spandex, micro-polyester and 100% polyester-knitted fabric', 'rib structure polyester-spandex, micro-polyester and 100% polyester-knitted fabric', 'plain structure polyester-spandex, micro-polyester and 100% polyester-knitted fabric'. Thickness and weight of specimen were influenced by density change caused by reducing and increasing fabric size. Thus, high density caused by floating in course-wise causes more knitted fabric weight gain than by loop overlapping.

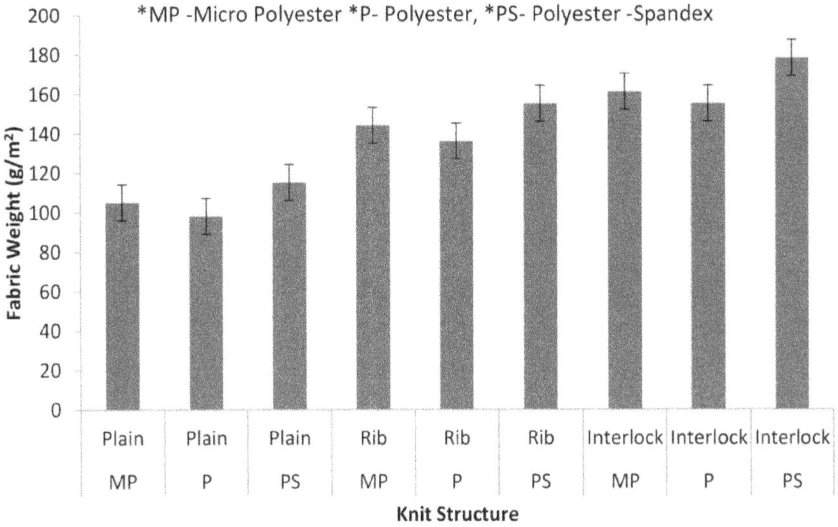

Figure 3.
Fabric weight comparison on knit structure.

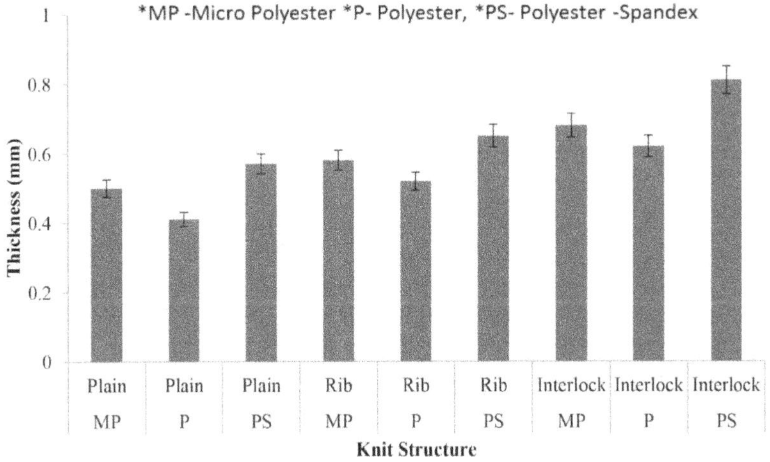

Figure 4.
Fabric thickness comparison on knit structure.

3.1 Stretch properties

Figure 5 and **Tables 2** and **3** indicate that the stretch value decreased in rib and interlock structure-knitted fabrics and direction except for wale-wise and course-wise as compared with plain structure fabrics. The plain structure fabrics have higher-stretch (%) polyester-spandex blend because lycra filament yarn have more stretch properties compared with other polyester and micro-polyester yarn [1].

The interlock structure three-knitted fabric showed a sharp decrease, while rib interlock structure three-knitted fabric had relatively small decrease. It seems that the material effect by stretch properties added to the reducing cause by yarn floating in the fabric structure, which held the loops reduced the stretch value of the fabric. The stretch value in course-wise is influenced by yarn floating rather than loop overlapping, while stretch value in wale-wise is caused by loop overlapping versus yarn floating [10].

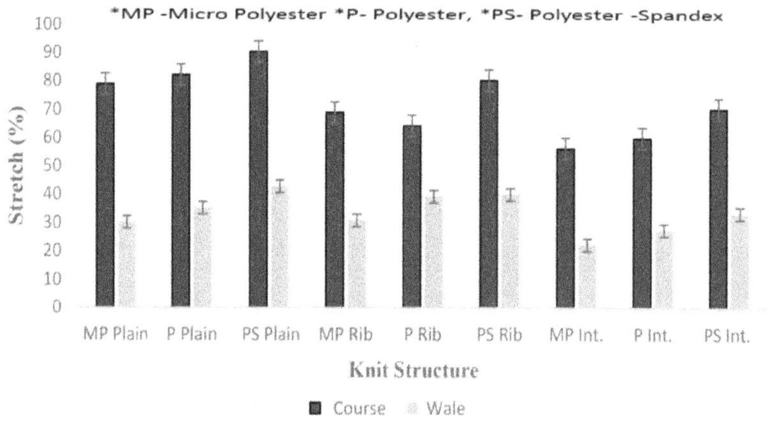

Figure 5.
Stretch comparison on knit structure.

Type of fabric	Structure	Fabric stretch and recovery properties					
		A	B	C	D	E	F
Polyester-spandex blende knitted	Plain	90.28	55.71	64.78	42.71	28.14	34.87
Polyester-spandex blende knitted	Interlock	70.14	29.07	35.85	33.25	19.85	20.11
Polyester-spandex blende knitted	Rib	80.21	38.57	51.21	39.51	22.42	27.31
Micro-polyester knitted	Plain	78.85	45.21	47.85	30.12	16.42	17.91
Micro-polyester knitted	Interlock	56.21	20.22	25.56	22.14	12.34	12.87
Micro-polyester knitted	Rib	64.85	33.28	35.61	30.81	14.78	16.83
100% Polyester knitted	Plain	82.12	52.27	55.22	35.14	20.15	25.12
100% Polyester knitted	Interlock	59.85	25.12	31.09	27.37	13.57	14.85
100% Polyester knitted	Rib	68.24	35.41	39.17	28.15	18.89	22.01

Note: A —*Course-wise stretch percentage,* **B** —*course-wise recovery after 60 sec %,* **C**—*course-wise recovery after 1 hr. %,* **D**—*Wale-wise stretch percentage,* **E**—*wale-wise recovery after 60 sec %,* **F**—*wale-wise recovery after 1 hr. %.*

Table 2.
Mean value of stretch and recovery test results.

Stretch properties	Degree of freedom df	Sum of square value SS	Mean square value MS	F_{actual}	$F_{critical}$	*P* value
A	9	5019.4	557.7	5.413	2.261	0.000
B	9	5728.4	636.4	3.167	2.261	0.006
C	9	6006.0	667.3	2.910	2.261	0.010
D	9	1284.1	385.5	2.813	2.261	0.013
E	9	1082.1	140.33	2.442	2.261	0.040
F	9	2205.2	245.03	5.536	2.261	0.023

Note: A—*Course-wise stretch percentage,* **B**—*course-wise recovery after 60 sec %,* **C**—*course-wise recovery after 1 hr. %,* **D**—*wale-wise stretch percentage,* **E**—*wale-wise recovery after 60 sec %,* **F**—*wale-wise recovery after 1 hr. %.*

Table 3.
One-way ANOVA of stretch and recovery properties of sportswear-knitted fabric structures.

Table 3 shows the ANOVA statistical analysis results at 5% significance level. Stretch and elastic recovery properties of the sportswear-knitted fabrics show significant difference between them (course-wise stretch (%): F_{actual} = 5.432 and wale-wise stretch (%): F_{actual} = 2.813 in comparison with $F_{critical}$ = 2.26) at degree of freedom 9.

3.2 Elastic recovery properties

There was significant value change on knit structure and direction in elastic recovery as shown in **Figures 6** and 7. The recovery value gap among knitted specimen was lower at 1 h than at 60 sec. The stretch loops bent and restricted by the external force loop of stretch take on a form of stability and shape retention in cover time [11–13].

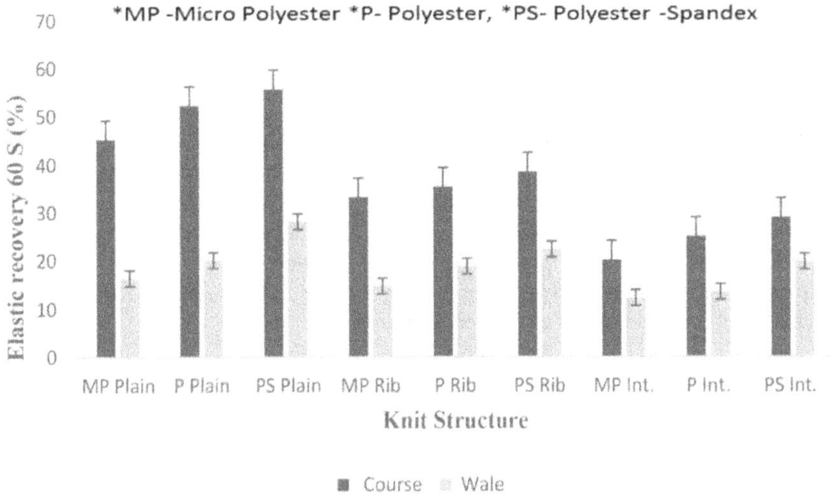

Figure 6.
Elastic recovery 60 sec on knit structure fabric.

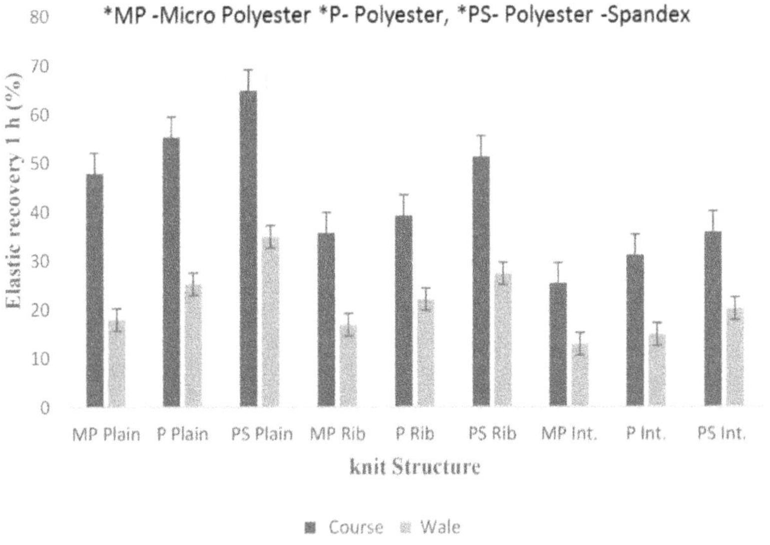

Figure 7.
Elastic recovery 1 h on knit structure fabric.

The ANOVA results show in **Table 3** that with respect to stretch properties after 60 sec %, there is a significant difference between the knitted fabric course-wise recovery after 60 sec %, degree of freedom 9. [$F_{actual} = 3.16 > F_{critical} = 2.26$ ($p < 0.05$)]. And for wale-wise recovery after 60 sec %, there is a significant difference between the structures [$F_{actual} = 2.44 > F_{critical} = 2.26$ ($p < 0.05$)].

It was found from **Table 3**, ANOVA results show that there is a significant difference between the course-wise recovery after 1 h %, value of knitted fabrics [$F_{actual} = 2.91 > F_{critical} = 2.26$ ($p < 0.05$)]. Also, it is noticed that there is a significant difference in wale-wise recovery after 1 h %, between the knitted fabrics [$F_{actual} = 5.53 > F_{critical} = 2.26$ ($p < 0.05$)].

4. Conclusion

The followings conclusions are derived from the above experimental work and given below:

- The elastic and recovery sportswear apparels are basically connected to the fabric material interface with the body, and these basically depend on how material stretch and recovery performance apparel structure perform. In this research study, basically focus should be on for the apparel size, stretch, elasticity of material properties. Apparel cloths have compression-knitted properties with different structures such as plain, interlock and rib and have analysed the correlation with type of materials.

- It is concluded that the plain structure polyester-spandex blend fabric is preferable than micro-polyester and 100% polyester fabric with respect to stretch and elastic recovery characteristics due to its quick recovery, which enhances the power of the performance.

- This chapter proposed an appropriate knit structure and arrangement approach in consideration of fabric size and stretch properties of high-stretch knitted fabric and correlation with type of fibre.

- This chapter provides meaningful market data for the effective development of more diverse garment-related product along with the localization of manufacturing for functional and sports garment.

Author details

Ramratan Guru[1*], Rajeev Kumar Varshney[2] and Rohit Kumar[2]

1 Department of Handloom and Textile Technology, Indian Institute of Handloom Technology, Varanasi, UP, India

2 Department of Textile Engineering, Giani Zail Singh Campus College of Engineering and Technology, MRSPTU, Bathinda, Punjab, India

*Address all correspondence to: ramratan333@gmail.com

References

[1] Senthilkumar M, Anbumani N. Dynamics of elastic knitted fabrics for sportswear. Journal of Industrial Textiles. 2011;**41**(1):13-24

[2] Rhie J. Fundamental relationship between extensibility of stretch fabric and its pressure. Family and Environment Research. 1992;**30**(1):1-2

[3] Chakraborty JN, Deora D. Functional and Interactive Sportswear. Asian Textile Journal. 2013;**22**(9):69

[4] Ramratan, Choudhary AK. Thermo-physiological study of active knitted sportswear: A critical review. Asian Textile Journal. 2018;**27**(7):53

[5] Ramratan, Choudhary AK. Influence of functional finishes on characteristics of knitted sportswear fabrics. Asian Textile Journal. 2018;**27**(8):43

[6] Yamada T, Matsuo M. Clothing pressure of knitted fabrics estimated in relation to tensile load under extension and recovery processes by simultaneous measurements. Textile Research Journal. 2009;**79**(11):1033

[7] Sang JS, Park MJ. Knit structure and properties of high stretch compression garments. Textile Science and Engineering. 2013;**50**(6):359-365

[8] Lyle D. Performance of Textiles. New York: Wiley and Sons, Inc.; 1977. pp. 168-169

[9] Senthilkumar R, Sundaresan S. Textiles in Sports and Leisure. The Indian Textile Journal. 2013;**123**(5):89-95

[10] Ladumor HC, Manish B, Vaishali DS. Elastic recovery characteristics of waist band using high stretch polyester in place of Lycra a technical review. International Journal for Scientific Research & Development. 2015;**3**(4):104-107

[11] Kentaro K, Takayuki O. Stretch properties of weft knitted fabrics. Journal of the Textile Machinery Society of Japan. 1996;**19**(4):112-117

[12] Senthilkumar M, Anbumanl N. Effect on laundering on dynamic elastic behavior of cotton and cotton spandex knitted fabrics. Journal of Textile And Apparel Technology and Management. 2012;7(4):1-10

[13] Ashayeri E, Alam FMS. Factors influencing the effectiveness of compression garments used in sports. Procedia Engineering. 2010;**2**(1): 2823-2829

[14] Su CI, Yang HY. Structure and elasticity of fine electrometric yarns. Textile Research Journal. 2004;**74**(12): 1041

[15] Song G. Improving Comfort in Clothing. Cambridge: Woodhead Publishing Limited; 2011. p. 114

[16] Shishoo R. Textiles in Sport. Cambridge, England: Woodhead Publishing in Textiles; 2005. pp. 1-8

[17] Robert SH, Fletcher HM. Elastic properties of plain and double-knit cotton fabrics. Textile Research Journal. 1964;**649**

[18] Saricam C. Absorption, wicking and drying characteristics of compression garments. Journal of Engineered Fibres and Fabrics. 2015;**10**(30):146-154

[19] Manshahia M, Das A. High active sportswear a critical review. Indian Journal of Fiber and Textile Research. 2014;**39**(2):441-449

[20] Morton WE, Hearle JWS. Physical Properties of Textiles Fibres. England: The Textile Institute Woodhead Publishing Limited; 1993

Chapter 9

Characterizing Stress-Strain Behavior of Materials through Nanoindentation

Indrani Sen and S. Sujith Kumar

Abstract

Nanoindentation is a widely used state of the art facility to precisely and conveniently evaluate the mechanical properties of a wide group of materials. Along with the determination of elastic modulus and hardness of materials, this chapter particularly aims to explore the possibilities to assess the corresponding stress–strain characteristics of elastic–plastic materials and most importantly unique pseudoelastic materials. The suitability of continuous stiffness measurement (CSM) based nanoindenter systems along with the adaptability of the instrument without CSM for precisely evaluating the deformation behavior of specialized materials is discussed in details. In this regard, the roll of indenter tip geometry and size is greatly emphasized. The recent research in the field is reviewed thoroughly and the updated protocol generated is illustrated.

Keywords: nanoindentation, stress-strain curve, small-scale, plasticity, NiTi

1. Introduction

Since the early 19th century, indentation technique has been extensively used for characterizing the mechanical properties of vast range of materials. In general, the indentation test is known to measure the *hardness* of materials. In conventional techniques, the mean contact pressure (*MCP*) upon indenting a specimen surface is evaluated. This is done on the basis of the residual area measured from the image of the indent impression and the known value of the applied load. The quantitative parameter, thus evaluated, represents the material's response against deformation. In fact, *MCP* measured at the fully developed plastic zone is known as hardness [1]. With the progress in the technology and its incorporation in the experimental setup, instrumented indentation technique, particularly '*nanoindentation*' has been evolved to assess various mechanical as well as metallurgical properties of a range of materials [2–4]. This includes characterizing elastic moduli, residual stress, creep properties, dislocation density, strain rate sensitivity etc. [5–12]. Among all these developments, the potential of the nanoindentation technique in generating the indention stress (σ_{ind}) – indentation strain (ε_{ind}) curve is the most recent one and it is explained in detail in this present chapter [2, 3].

In nanoindentation, the associated high-resolution depth sensing technique aids to estimate the depth or size of the deformation zone. The process records the continuous response of indentation load (P) in the range of μN vs. indentation depth (h) in the magnitude of nm. The P-h curve obtained therein helps to assess the various properties of the studied materials. Unlike the conventional technique, in instrumented nanoindentation, hardness is estimated by using indirect measurement of projected contact area from P-h curve and the known geometry of the indenter tip. Similarly, elastic modulus of the material is estimated using the slope of the unloading segment in the P-h response of materials [13]. This method of analysis has been used for various scientific studies to characterize the localized mechanical properties of the samples in sub-micron scale. In fact, this revolutionary modification in the assessment methodology through nanoindentation has opened up a wide range of studies to extract the different relevant mechanical properties of materials on a small-scale.

One of the breakthroughs is the capability of this technique in generating the σ_{ind} - ε_{ind} response of a material of interest [3, 14–16]. This novel and recent development plays a significant role in understanding the localized deformation capability of materials system. This is particularly because stress – strain characteristics can provide an insight into the elastic – plastic mechanisms of the materials, as per the conventional notion. In fact, estimation of localized stress – strain characteristics of a material through nanoindentation can even be a substitute for typically used small-scale characterization techniques for instance, micro-pillar compression [17, 18]. Nevertheless, nanoindentation is further beneficial owing to its easier sample preparation, simplicity in experimental execution, and non-destructive nature. This technique therefore has enormous potential for evaluation of small-scale mechanical properties of materials with minimal effort.

Considering this, the present chapter is dedicated to provide a reasonable understanding for generating σ_{ind} - ε_{ind} data from the P-h curve of nanoindentation. To develop a more conceptual idea for a new reader, the importance of indenter tip geometry in activating different deformation modes within the indented volume are discussed at the first hand. Subsequently, the basic relationships for the indentation, the method of analysis and generation of protocol for obtaining the σ_{ind} - ε_{ind} curve will be discussed.

2. Role of indenter configuration

It is noteworthy that both the uni-axial tensile/compression test as well as the indentation technique are capable to assess the stress–strain characteristics of a material, however, with usually different size-scale of samples along with varying stress-states. The former provides an understanding for the degree of bond stretching induced elastic deformation and dislocation mediated plastic/permanent deformation in the material. To obtain such desired information, the strain-induced into the material should be controlled in such a way that, the material's response reflects the gradual activation and transition from the elastic to the plastic deformation. This is realized in uni-axial deformation without any strain gradient in the specimen, at least macroscopically.

In contrary to that, upon indentation, presence multi-axial state of stress exists beneath the indenter tip. Moreover, the constraint nature of deformation induces strain gradient within the deformation volume. Hence, for assessing the elastic–plastic activity within the deformation zone, the indentation tests need to be specially designed to produce a smooth strain distribution (or gradient) along with its gradual increment. To maintain that, indenter tip geometry needs to be carefully chosen to reflect the σ_{ind} - ε_{ind} characteristics from the localized region. In this regard, the most suitable indenter configuration is spherical tip (or sphero-conical indenter).

Before getting into the details about the configuration of the spherical indenter tip and its importance for σ_{ind} - ε_{ind} generation, the reader needs to develop a comprehensive idea about the different type of indenter tips that are used in general. For the same, the geometrical aspect of indenter configuration is briefed here. From a geometrical point of view, indenters are classified into two: (i) geometrically similar indenters (*GSI*) and (ii) non-geometrically similar indenter (*N-GSI*) [4].

The most commonly used sharp pyramidal indenter such as four-sided Vickers (for micro-and macro-indentation) and three-sided Berkovich indenters (for nanoindentation) comes under the category of *GSI*. On the other hand, the spherical indenter falls under the category of *N-GSI*. The major difference in deformation characteristics experienced by a specimen surface, by indenting with any of these two categories of indenter tips can better be appreciated from **Figure 1**. Schematic representations in **Figure 1(a-c)** show the deformation modes activated in traditional elastic–plastic material while increasing the indentation load/depth, using *GSI*. The mathematical relation for geometrical similarity originates from the ratio of the contact radius (a_c) to the maximum depth of indentation (h_{max}). For *GSI*, a_{c-i}/h_{max-i} = a_{c-j}/h_{max-j} = a_{c-k}/h_{max-k} = constant. The subscript, i, j, k signify increasing level of h. Nevertheless, this constant ratio of a_c/h_{max} ensures that the size of the deformation zone of indentation varies uniformly irrespective of the depth of penetration. This helps to estimate the property of the subjected material independent of the applied indentation load/depth. Nevertheless, owing to the sharp nature of the Vickers and Berkovich indenter, the strain-induced within the indentation volume is large enough to generate significant plastic deformation [1]. In that case, dislocation activity is always the dominant mechanism within the deformation volume beneath the indenter tip, irrespective of the change in depth of indentation, as apparent from **Figure 1**. This assists to precisely measure the hardness of a material independent of the indentation load, in the theoretical sense. However, it is realized that *GSI* is not adequate to assess the elastic deformation response of the indented material. In fact, while using conventional Vickers and Berkovich

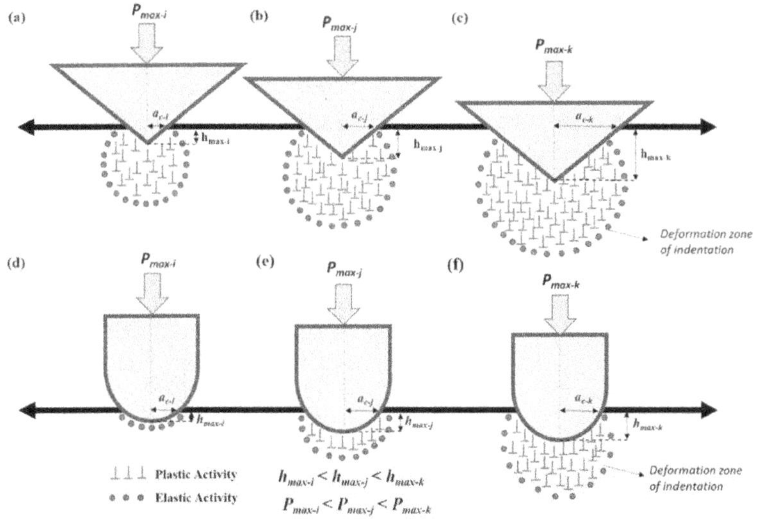

Figure 1.
Schematic illustration of the indentation behavior associated with traditional elastic-plastic metallic using (a-c) sharp geometrically similar indenter and (d-e) spherical non-geometrically similar indenter at various indentation depths.

indenters, occurrence of prominent dislocation activity within the deformation volume negates any influence of elastic activity therein. This acts as the limitation of the most commonly using Vickers and Berkovich indenter for generating the stress–strain curve.

On the other hand, a completely different deformation response is being experienced, while the specimen surface is indented using *N-GSI* (spherical tip) with increasing indentation load/depth. **Figure 1(d-f)**, illustrate the deformation scenario within the indentation volume, in such case. It is evident from the figure that, nature of deformation is entirely different in comparison to that for *GSI*. This difference originates from the non-geometrical similarity of the indenter. In case, the specimen surface is indented with a *N-GSI*, $a_{c\text{-}i}/h_{max\text{-}i} < a_{c\text{-}j}/h_{max\text{-}j} < a_{c\text{-}k}/h_{max\text{-}k}$. This essentially means with the progress of the indentation, increment in the contact radius becomes more pronounced with respect to the depth of penetration. Such movement of indenter within the material surface gradually increases the induced strain/stress into the material. Also, the blunt nature of the indenter assists in generating a smooth stress field within the indentation volume, specifically as compared to *GSI*. As a net effect, spherical indenter facilitates a gradual activation of elastic to the plastic deformation mechanism. This potential for gradual instigation of the deformation mechanism similar to that observed in case of uni-axial test, is exploited for σ_{ind} - ε_{ind} generation from nanoindentation.

Nevertheless, the most crucial part in this regard is the data analysis procedure that is necessary to convert the indentation *P-h* response into a reliable σ_{ind} - ε_{ind} curve. There have been numerous attempts to obtain a stress–strain curve from traditional indentation as well as instrumented one. In the process, the protocol for generation of indentation stress–strain curve has undergone various alterations, to precisely correlate the materials' property. In the next section, we have briefed the different approaches adopted to appreciate the σ_{ind} - ε_{ind} behavior of a material. This will help to understand the scientific developments that has been materialized on this particular topic, so far.

3. Evolution of σ_{ind} - ε_{ind} generation protocols

The concept for the generation of σ_{ind} - ε_{ind} curve from indentation is introduced by Tabor in the 1950s. Tabor has measured the *MCP* on the specimen indented with a spherical tip to estimate the stress that is induced in the process [1]. The most crucial part, however, is the estimation of ε_{ind}. Tabor defined ε_{ind} by the relation (d/D), where d is the diameter of the residual impression and D is the diameter of the indenter tip. Here d is measured using the traditional approach, i.e., by imaging of residual impression after unloading. The general trend of σ_{ind} - ε_{ind} characteristics of materials, generated following Tabor's protocol, resembles well with that evaluated through traditional uniaxial compression test [1]. However, this method of analysis accounts for only single σ_{ind} - ε_{ind} data from an indentation. So, it means that several indentation tests with different indentation parameters are necessary to be pursued, to obtain a continuous σ_{ind} - ε_{ind} curve for a material, making the process cumbersome.

Nevertheless, Tabor's approach revealed the potential of the indentation technique and instigated more studies to develop a state-of-the-art protocol for generating σ_{ind} - ε_{ind} curve of a material. In this regard, automation through the instrumented indentation has opened up enormous possibilities to generate the σ_{ind} - ε_{ind} curve using a single indentation. In turn, the localized deformation behavior of a material can be precisely obtained. First among all is the Field and Swan approach [19]. They have proposed to incorporate multiple partial unload segments during each indentation. Here, the *P-h* responses obtained for each particular segments are used to measure the corresponding σ_{ind} and ε_{ind} values. The strain, on the other hand, is estimated using the relation a/R_i, where R_i represents the radius of the indenter tip. As per Field and Swan approach, the deformation associated in each unloading segment is assumed to be purely elastic. Correspondingly, the classical Hertzian elastic relationship (explained in the next section by Eq. (1)) is applied on those *P-h* responses to assess the contact radius, a. From the measured a value, contact area (A_c) is estimated instead of residual impression-based analysis in Tabor's protocol.

The Field and Swan approach has much significance in the present scenario, owing to its implementation of the Hertzian contact mechanics theory. Nevertheless, interpretation of indentation strain as per both Tabor's as well as Field and Swan approaches has been questioned for its integrity with the fundamental concept of strain. In general, strain is defined as the ratio of change in length to the initial length in a region of deformation considered. However, this fundamental relationship is not met in both these above-mentioned approaches.

In order to overcome this fundamental lacking, various studies have been conducted to formulate an adequate relationship for the ε_{ind}. Among those attempts, the protocol developed by *Kalidindi* and *Pathak* has succeeded in defining ε_{ind} as per the most basic concept of strain [16]. The present chapter is extensively covering the formulation and implementation of *Kalidindi* and *Pathak* protocol for the generation of σ_{ind} - ε_{ind} curve for a material subjected to nanoindentation. This protocol is essentially formulated based on classical Hertzian theory, which is explained below.

4. Contact mechanics for spherical tip-based indentation

Contact mechanics theory introduced by Hertz has provided a fundamental basis for the indentation technique [20]. Classical Hertzian theory predicts the elastic responses of frictionless contact between two different bodies of dissimilar geometries (with varying properties) in contact. This theory is formulated based on the assumption that material is homogenous and isotropic. In the present scenario of indentation using spherical indenter, the Hertzian theory for elastic contact between the sphere (indenter) and elastic half-space (specimen surface) is used for the formulation of σ_{ind} - ε_{ind} generation. In the indentation aspect, the material of interest is considered as an elastic half-space by following the criteria that indenter tip radius (R_i) should be at least ten times smaller than the horizontal dimensions of the sample [21].

As explained in previous Section 2 (see **Figure 1(d-f)**), indentation using spherical indenter tip facilitates the gradual activation of elastic to plastic mechanisms in the material. Therefore, for the sake of understanding, the overall deformation scenario can be categorized into (i) fully elastic and (ii) plastic following the initial elastic section. The schematic representation of these two modes of deformation and their corresponding *P-h* response is showed in **Figure 2**. In the first case, material recovers all the depth it penetrated upon the indentation (see **Figure 2(a)** and **(c)**). In the second case, some amount of permanent deformation is existing within the indentation volume (see **Figure 2(b)** and **(d)**). Hertz has provided the basis for the elastic deformation associated in two former cases using the relation below,

$$P = \frac{4}{3} E_{eff} R_{eff}^{1/2} h_r^{3/2} \tag{1}$$

$$\frac{1}{E_{eff}} = \frac{1-\nu_s^2}{E_s} + \frac{1-\nu_i^2}{E_i}, \quad \frac{1}{R_{eff}} = \frac{1}{R_i} - \frac{1}{R_s} \tag{2}$$

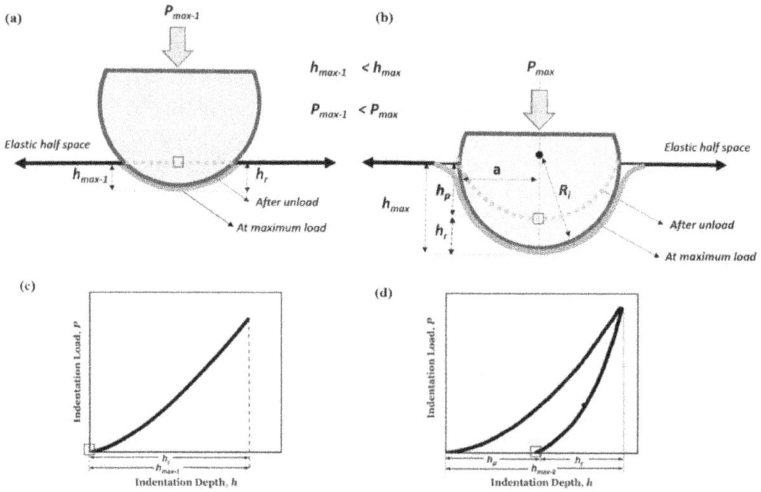

Figure 2.
Schematic representation of indentation of behavior of material in (a) fully elastic condition and in the pressure of (b) plastic deformation. Corresponding indentation load vs. indentation depth responses of materials are shown in (c) and (d).

Here P is the applied load, h_r is the recoverable depth, R_{eff} is the effective tip radius and E_{eff} is the effective elastic moduli. All the characteristic terms mentioned here can be appreciated from **Figure 2(b)**. The terms h_{max} and h_p in **Figure 2** represent the maximum depth of indentation at P_{max} and recurring plastic depth of indentation post-unloading (P is zero), respectively. In the Hertzian relation, the role of elastic deformation on the two mating parts is assessed using E_{eff}. The value of E_{eff} accommodates the elastic deformation associated with the hard indenter and soft sample. E_{eff} during the indentation is estimated using the relation (2). Similarly, R_{eff} takes into account the influence of plastic activity on the overall deformation. It is related to the indenter tip radius (R_i) and the radius of curvature of the sample (R_s) upon the indentation. R_{eff} of the sample is estimated using the relation (2).

All these relations derived by Hertz has laid the foundation for the formulation of σ_{ind} - ε_{ind} data from the nanoindentation P-h response. This is explained in details in the following section.

5. Defining the indentation stress and indentation strain

It is well understood from Section 3 that Tabor's and Field and Swan's protocols do not suffice to define the ε_{ind} precisely. Nevertheless, *Kalidindi* and *Pathak* have defined the σ_{ind} and ε_{ind} by considering the size of the deformation zone formed beneath the indenter and correlated it with the fundamental Hertzian relationship [16]. This protocol has succeeded in producing comprehensive σ_{ind} - ε_{ind} data from the nanoindentation experiments (explained in Section 6).

As per this novel approach, eq. (1) is rearranged by incorporating the following relations:

$$\sigma_{ind} = \frac{P}{\pi a^2}; \sigma_{ind} = E_{eff}\varepsilon_{ind}; \varepsilon_{ind} = \frac{4}{3\pi}\frac{h_r}{a} \approx \frac{h_r}{2.4\,a} \quad (3)$$

$$a = \sqrt{R_{eff}h_r} \quad (4)$$

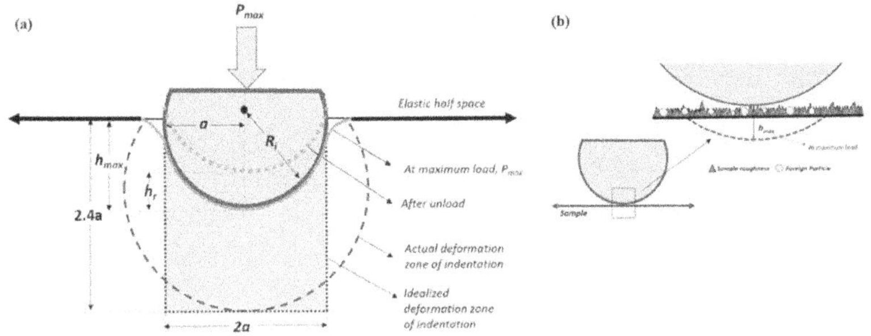

Figure 3.
(a) Schematic representation of the deformation behaviour associated with indentation. Figure highlights the actual deformation zone of indentation and the idealized deformation zone of indentation. (b) Schematic representation of surface irregularities on a sample.

The indentation strain defined using the above relationship satisfies the general definition of strain. This can be better appreciated from **Figure 3(a)**. In the figure, the dashed spherical shaped region beneath the indenter tip schematically shows the actual size of the deformation zone upon indentation. Based on the ε_{ind} defined from Hertzian relation, the length of the deformation zone beneath the indenter tip at P_{max} is noted to be $\sim 2.4a$. Interestingly, a simulative study on the prediction of indentation behavior strongly agrees with this relation for tungsten and aluminium [16]. This has validated the new definition of ε_{ind}, which is derived without any alteration of the fundamental Hertzian relation. This novel protocol is remarkably different yet comprehensive with respect to the other discussed approaches. This is primarily because it basically takes into account the actual size of the deformation zone during the indentation, rather than simply estimating the ε_{ind} data using the concept of variation in indent impression.

Furthermore, this novel protocol has provided a reasonable basis for the analogical comparison of indentation behavior using spherical indenter and uniaxial compression test. The overall nature of the material response upon nanoindentation can be considered as the replication of compressing up to a depth of h_{max} on a cylindrical sample of height $2.4a$ and radius a. To visualize it clearly, the idealized deformation zone of indentation and actual deformation of indentation is schematically shown in **Figure 3(a)**. The shape of the actual deformation zone formed is schematically showed as spherical. The reader should be aware that, in reality, owing to the anisotropy in material's properties, the actual shape of the deformation zone of indentation can be slightly different from this schematic representation. It is also noteworthy that with slight alternation in relation (4), h_{max} can be used instead of h_r in the numerator to accommodate the plastic activity [15]. This whole theoretical concept has paved the way for generating σ_{ind} - ε_{ind} curve from the *P-h* signal in nanoindentation. To realize it in a practical scenario, the reader has to understand the necessary steps to follow for obtaining a reliable output.

6. Theoretical conceptualization to experimental execution

As mentioned in Section 1, nanoindentation typically generates a *P-h* response and its characteristics define the mechanical property of the material indented. Compared to any other characterization technique, particularly, the most commonly used uni-axial tests, the size of the active deformation region for nanoindentation is extremely small. Therefore, proper measures are necessary at every steps right from the precise sample preparation to the careful data analysis to obtain reliable data.

6.1 Sample preparation

The existence of an artefacts such as scratches or the presence of foreign particles on the surface can influence the *P-h* signal and thereby the generated σ_{ind} - ε_{ind} data. The poorly polished samples create a scratch on the surface, the depth of which can be in hundreds of nanometres. Data recorded from such a region will certainly influence the overall σ_{ind} - ε_{ind} characteristics and consequently alter the assessment of the true properties of the material. This can be visualized and understood from the schematic representation in **Figure 3(b)**. In the figure, red coloured triangular shape and yellow coloured circular shape reveal the presence of sample surface roughness and foreign particles respectively. As per the indentation sequence, the indenter will first acquire the data from those artefacts and move to the bulk of the sample. So, actual material which is supposed to show the pure elastic response initially, is now influenced by the presence of sample surface artefacts. As a net effect, the *P-h* response from the bulk sample is influenced by the surface roughness/foreign particle. Hence, the assessed properties are certainly different from the true ones [1]. In case of conventional uniaxial tests, such misinterpretation of results can be obtained in case a specimen slips upon loading, or even when elastic properties are estimated from a tensile experiment, without attaching an extensometer to the test specimen.

To avoid such issues, well-polished, smooth, flat and plane-parallel specimen should be subjected to nanoindentation. The necessary steps to achieve such artefact free surface vary with the material of interest. However, colloidal silica polish for few hours (minimum 3 h) after the conventional polish using silicon carbide paper with decreasing mesh size and diamond polish is prescribed for metallic specimens, to attain a reasonably good surface condition for the σ_{ind} - ε_{ind} generation. Depending on the surface characteristics of the material, electropolishing may also appear to be a better option to minimize the artefacts on the sample surface.

6.2 Conversion of experimental *P-h* data to effective *P-h* data

It is noted that theoretical predictions and the experimental outcome may result to some disparities in case of the nanoindentation test. In this regard, it is noteworthy that proper data analysis plays a key role in the generation of σ_{ind} - ε_{ind} curve. It is highlighted in the previous section (Section 6.1) that nanoindentation experiments mandate extremely good quality surface finish. Nevertheless, obtaining the required surface finish is difficult in practice. A proper data correction route on the experimentally obtained *P-h* curve, on the other hand, can negate the role of artefacts on the σ_{ind} - ε_{ind} analysis. This step is crucial to compute a reliable stress–strain curve. For the same, effective initial contact point between the indenter tip and the specimen surface is estimated following the "zero-point correction" (*ZPC*). In fact, *ZPC* deals with discarding the data points which are influenced by unavoidable surface irregularities. In turn, the effective contact point is determined on the basis of Hertzian theory which reciprocates the material behavior. According to the type of nanoindentation instrument used, *Kalidindi* and *Pathak* have proposed two different approaches for the data correction using *ZPC*. One is for nanoindenter with (a) Contact Stiffness Mode, *CSM* (or Dynamic Mechanical Analysis, *DMA*) and another for (b) Non-Contact Stiffness Mode, *N-CSM* [22]. These two modes are slightly different in the method of experimentation.

6.2.1 CSM *mode or* DMA *mode*

In *CSM* or *DMA* mode, harmonic force is imposed in the loading and unloading segment during the indentation. This is highlighted at the inset (a) of **Figure 4**. It can be hypothetically viewed as if the specimen undergoes multiple indentations with minimal depth scale (2 to 4 nm) while conducting a single indentation. Displacement responses corresponding to these harmonic forces are recorded throughout the indentation. These assist in assessing the variation in contact stiffness, S (*or* $\frac{dP}{dh}$) upon the indentation. Precise determination of S from each steps of *CSM* leads to estimate the continuous variation in the related properties of materials with increasing h, for example, hardness and elastic modulus changes [22].

In the present scenario, the continuously varying S, h_r, and P are obtained from the *CSM* mode of the nanoindenter and these signals are used for *ZPC*. For the same, the Hertzian relation (Eq. (1)) for elastic contact is rearranged into the following relationship,

$$P - \frac{2}{3}h_r S = -\frac{2}{3}h^* S + P^* \tag{5}$$

Here P^* and h^* denote the effective indentation load and depth respectively. A linear regression analysis on relation (5) helps to trace the P^* and h^* values through the slope ($-\frac{2}{3}h^*$) and y-intercept (P^*). Once the P^* and h^* are established, the experimentally generated *P-h* signal has to be corrected for obtaining an '*effective P-h curve*', which is devoid of any influence from the surface artefacts [16].

*6.2.2 Non-*CSM *mode*

In *N-CSM* mode, indentation is performed without harmonic force. This is also highlighted in inset (b) of **Figure 4**. In this particular case, *ZPC* is performed by recasting the Hertzian equation as per the relation below (derived from Eq. (1),

$$(h_r - h^*) = k\,(P - P^*), k = \frac{3}{4}\frac{1}{E_{eff}}\frac{1}{\sqrt{R_{eff}}} \tag{6}$$

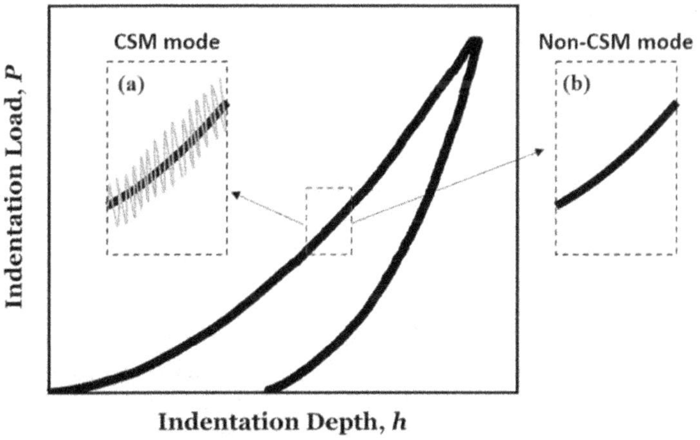

Figure 4.
Indentation load vs. indentation depth response generated using nanoindentation. Insets in the figure highlight the method of indentation in CSM mode and N-CSM mode.

In the above relationship, the k value is constant in the elastic segment [15]. It is worth reiterating here that within the elastic segment, continuously varying h equates with h_r whereas R_{eff} to R_i (explained in the subsequent Section 6.3). Also, prior understanding of elastic moduli of the material makes the calculation much easier. Essentially, regression analysis on the initial elastic segment of experimentally obtained data helps to calculate the values of P^* and h^* and thereby the effective *P-h* data is estimated.

6.2.3 Selection of data segment

The above-mentioned data correction procedures for nanoindenter with *CSM* or *N-CSM* mode, ideally has to be performed on the initial elastic segment of the *P-h* segment. Such elastic segment dwells within few nanometers, in reality. The exact value of this elastic segment however varies with the sharpness (or bluntness) of the indenter tip and the associated variation in strain gradient [2]. The question here is how to precisely choose a segment in the *P-h* curve which can be used for the data correction using Eqs. (5) and (6). This can be realized through the iteration process on the initial segment with a different depth limit. For instance, for nanoindentation with h_{max} of 250 nm, regression analysis has to be performed in initial segments with indentation depth of 10 nm (or any other limit) to higher. By doing so, the accurate point of transition from elastic to plastic (data limit) can be approximately finalized based on the continuity nature observed in the effective *P-h* curve as well as the corresponding σ_{ind} - ε_{ind} curve (explained in next Section 6.3).

6.3 Conversion of effective *P-h* curve to σ_{ind} - ε_{ind} curve

As explained in Section 5, the Hertzian relation has provided a basis to obtain σ_{ind} - ε_{ind} curve from the *P-h* response. Once the effective *P-h* response is computed using the steps mentioned in Section 6.2, Eqs. (3) and (4) are used for obtaining the corresponding σ_{ind} - ε_{ind} values. In this conversion process, estimating the continuously varying a is important for calculating the continuous evolution in the σ_{ind} and ε_{ind} values. It is particularly evident from Eq. (4) that, a is the main characterizing parameter to obtain the σ_{ind} and ε_{ind} values.

Prior to going through further details, the physical significance of a and the mechanisms behind its alteration during indentation are explained through **Figure 5**. The figure schematically shows the indentation behavior of different materials with different extents of elastic–plastic activities. Sample-1 with green color indicates the material with full elastic recovery. Sample-2 (orange) and sample-3 (blue) exhibit the indentation behavior of two materials with different degrees of plastic activities along with elastic deformation. In a fully elastic material (sample-1), the indented surface recovers the whole depth upon the complete removal of load. Thereby R_s attains infinity in this case (see **Figure 5**). So, $R_{eff} = R_i$ for material with full depth recovery (see equation (2)). Similarly, owing to the full recovery, continuously recording h signal can equate with the depth recovery (h_r). In short, $R_{eff} = R_i$ and $h = h_r$ within the elastic regime of material upon indentation.

But, once the dislocation mediated plastic activity is instigated, R_s attains a finite value. The orange and light blue colors in **Figure 5** reveal the formation of finite values of R_s in the materials due to the occurrence of plastic deformation. In these two cases, R_{eff} is no longer equal to R_i. It is reported that R_{eff} is significantly larger than R_i once plastic deformation initiates in the material. Almost a 100-fold increment in the R_{eff} is reported with presence of plastic activity in aluminum sample [3].

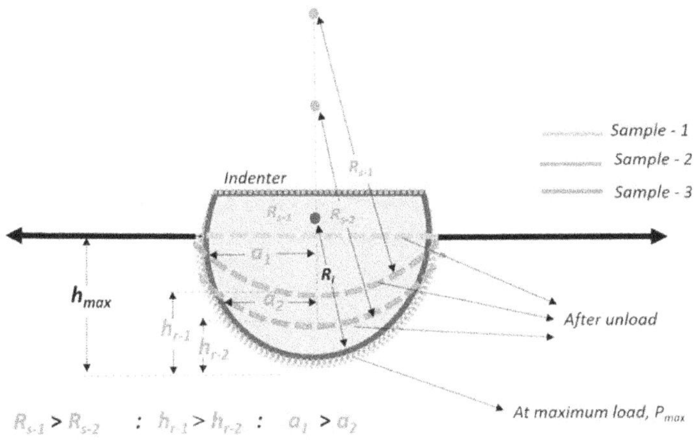

Figure 5.
Schematic representation of the nature of deformation volume beneath the indenter tip for materials with three different degree of elastic-plastic property. Green line shows the sample with full depth recovery. The orange and light blue colors reveal the indentation behaviour of samples with different shares of elastic and plastic activities.

All these physical changes are also related to h_r after the unloading. While comparing sample-2 and sample-3, depth recovery is noted to be higher for the former. Correspondingly, R_s in the material also changes. It is apparent from the **Figure 5** that $R_{s\text{-}1} > R_{s\text{-}2}$ and $h_{r\text{-}1} > h_{r\text{-}2}$. As a net effect of change in R_{eff} and h_r, contact between the indenter and sample deflects. This is reciprocated in the changes in a ($a_1 > a_2$). In conclusion, all three parameters are correlated which are primarily controlled by the share of elastic–plastic activities within the material of interest. Eq. (4) derived by Hertz relates all these physical phenomena and in the present scenario, it is utilized to estimate σ_{ind}-ε_{ind} curve using relation (3).

6.3.1 σ_{ind}-ε_{ind} from CSM *nanoindenter*

Estimation of a from nanoindentation using *CSM* mode is straight forward. The interrelation between S and a are derived from Eqs. (1) and (4) according to the Hertzian theory, as shown below:

$$\frac{dP}{dh} = 2\,E_{eff} R_{eff}^{1/2} h_e^{1/2} = 2\,E_{eff} a \qquad a = \frac{S}{2\,E_{eff}} \tag{7}$$

The analytical significance of this mathematical derivation lies in the fact that unlike Tabor's approach, this expression (eq. (7)) enables to assess the nature of deformation inside the material without a visual inspection of residual impression.

In the data analysis, once the evolving values of a are established using eq. (7), the final σ_{ind} - ε_{ind} curve is generated from Eq. (3). **Figure 6(a)** shows σ_{ind} - ε_{ind} curves obtained before and after *ZPC* on experimental *P-h* data. Interestingly, in this novel protocol, elastic moduli measured from the loading and unloading segments of the σ_{ind} - ε_{ind} curve are noted to resemble each other [16]. This observation has validated the new definition for ε_{ind} as well as the novel protocol for reliably assessing the mechanical property via σ_{ind} - ε_{ind} curve.

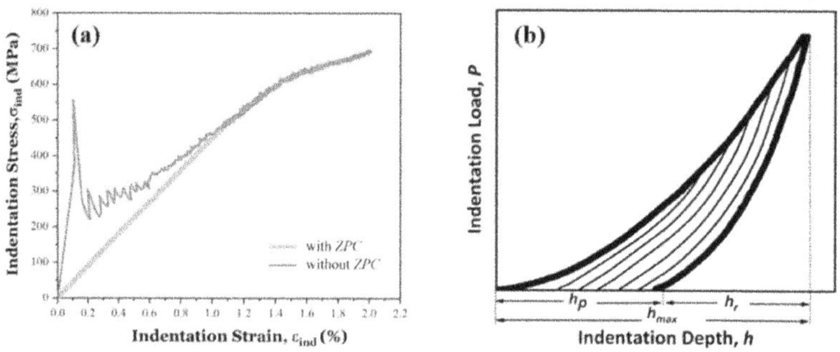

Figure 6.
(a) $\sigma_{ind} - \varepsilon_{ind}$ curve obtained before and after the zero-point correction. (b) Schematic representation of the P-h responses with multiple unloading segments for generating $\sigma_{ind} - \varepsilon_{ind}$ curve in N-CSM measurement nanoindenter.

6.3.2 σ_{ind}-ε_{ind} from *non-*CSM *nanoindenter*

As compared to the *CSM* mode, experimentation and method of analysis is different in case of *N-CSM* mode of nanoindentation. In *N-CSM* mode, multiple unloading segments are introduced into the indentation test for measuring the evolution in a and thereby the continuous variation in σ_{ind} and ε_{ind} values. This is similar to Field and Swan approach in terms of experimentation. **Figure 6(b)** schematically shows the *P-h* curve obtained after the multiple unloading. Once the effective *P-h* curve is generated by employing *ZPC*, following Section 6.2, the evolving values of a are estimated from each segment. For the same, R_{eff} value is estimated by fitting the unloading response using the modified Hertz relation as mentioned below,

$$h_r = h_{max} - h_p = k\,P^{2/3} \tag{8}$$

here k is a function of R_{eff} and E_{eff} (see relation (6)). E_{eff} can be traced from the prior understanding of elastic moduli of sample or from the initial elastic segment in the *P-h* curve [15]. So, from the understanding of k value of the respective alloys and the recorded value of h_p with reduction in indentation load in the unloading segments, R_{eff} is estimated by fitting using the relation (8). Once R_{eff} is established, a can be determined from relation (4) and in turn σ_{ind} - ε_{ind} curve can be generated using Eq. (3). It is also important to note here that, number of data points in the resultant σ_{ind} - ε_{ind} curve depends on the number of unloading segments provided in the experiment.

7. Protocol for σ_{ind} - ε_{ind} generation in pseudoelastic shape memory alloys

Previous sections have elaborated the potential of the nanoindentation technique in appreciating the σ_{ind} - ε_{ind} characteristics of traditional elastic–plastic metallic systems. In a further extension, *Sujith* and *Sen* have revealed the capability of nanoindentation in assessing the unique pseudoelastic (or superelastic) properties of shape memory alloys (*SMA*) *via* σ_{ind} - ε_{ind} curve [2, 6]. This recent development has succeeded in the producing the specialized stress - strain characteristics of the pseudoelastic NiTi system using most commonly used *N-CSM* nanoindenter.

It is noteworthy at this point that as compared to the traditional elastic–plastic metallic alloys, pseudoelastic system is different owing to the occurrence of reversible stress-induced martensitic transformation (*SIMT*). In pseudoelastic alloys (some examples of metallic systems are NiTi, Cu-Al-Zn, Cu-Al-Ni, Ni-Ti-Fe, Fe-Mn-Si, Fe-Mn-Si-Co-Ni), parent austenitic phase transforms to product martensitic phase upon the application of stress and it reverts to the previous austenite with the release of stress. Owing to this reversible *SIMT* along with usual elastic deformation in the parent and product phase, the NiTi system in pseudoelastic state shows (8–10) % of recoverable strain. This is also reflected as a unique characteristic in the conventional uni-axial stress - strain curve. Hence, evaluating such unique property using nanoindentation requires special attention in terms of (a) optimizing indentation parameters as well as (b) tailored σ_{ind} - ε_{ind} generation protocol. This investigation by *Sujith* and *Sen* is the first of its kind to consider spherical indenter tips with varying R_i as well as P_{max} levels with the aim to identify the optimum combination to precisely evaluate localized pseudoelasticity in *SMA* through nanoindentation. Following steps are briefed:

7.1 Optimizing indentation parameters

For optimizing the indentation parameters, a detailed analysis is performed on the *P-h* curve obtained from various indenter configuration (R_i of 10 μm, 20 μm and 50 μm) as well as P_{max} (1 mN to 7 mN). Details of the experiments and analysis procedures are reported elsewhere [2]. However, the key observations in this method of analysis are mentioned here.

Optimization of indentation parameters is performed based on the close scrutiny of the experimentally generated *P-h* curve using Hertzian theoretical prediction and the understanding of the pseudoelastic behavior in the alloy system. **Figure 7(a)** shows the method of analysis performed on the *P-h* curve. The black solid and the red dashed curves in **Figure 7(a)** show the experimental results and Hertzian theoretical predication of indentation response, respectively. Using this comparison, overall deformation mode in the indentation is parted into different sections. Correspondingly, the depth of indentation, specifically influenced by pseudoelasticity is assessed. This can be even better appreciated from **Figure 7(b)**. Physical variation associated with indentation volume of NiTi sample using R_i of 10 μm and 20 μm are schematically (in two halves) shown in **Figure 7(b)**. The region influenced by reversible *SIMT* is highlighted as green color in the schematics. This novel method of analysis is performed using a range of combination of indentation parameter. The most adequate combination to assess the pseudoelasticity is identified based on the share of reversible *SIMT* activity and the overall depth recoverability (minimum 90% depth recovery). Based on this systematic analysis, spherical indenter with R_i of 20 μm and P_{max} of 5 mN is noted to be most suitable combination for appreciating pseudoelasticity devoid of the influence of dominant plasticity, in NiTi system.

7.2 σ_{ind} - ε_{ind} protocol

Considering the extremely high depth recoverability (≥ 90%) of pseudoelastic NiTi system, following assumption is used while generating the corresponding σ_{ind} - ε_{ind} curves,

$$R_{eff} = R_i \text{ and } h_r = h \tag{9}$$

Section 6.3 has already mentioned about the validity of this assumption when material shows full depth recovery. In the present scenario, same assumption is used with depth recovery limit of 90% of the h_{max}. This assists in converting the *P-h* response into σ_{ind} - ε_{ind} curve using Eq. (3), while employing relations (9) in it. Essentially, this new protocol defined the ε_{ind} and a using following relation,

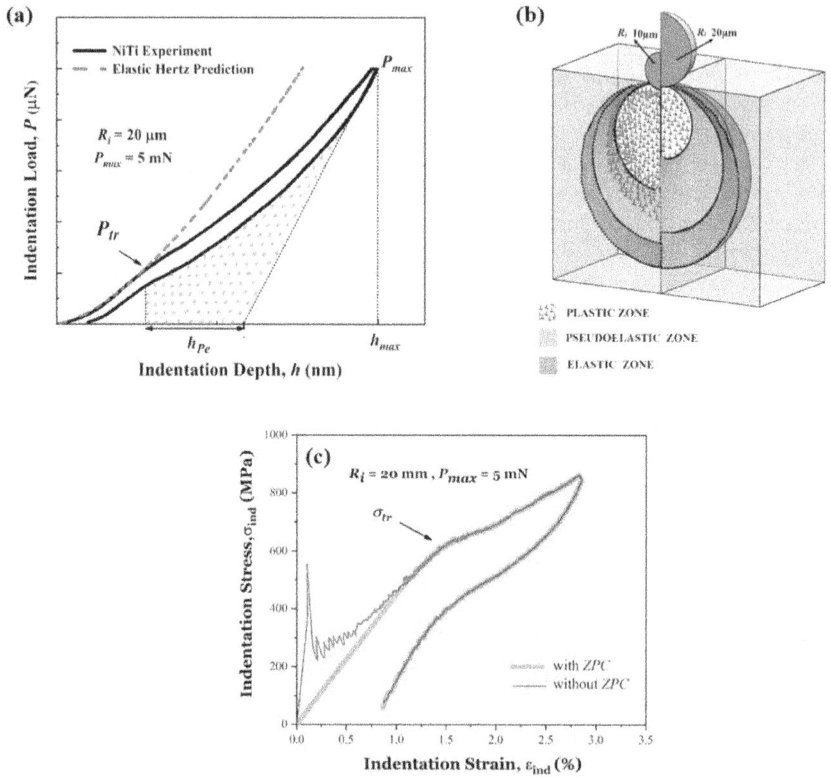

Figure 7.
*(a) The P-h response of pseudoelastic NiTi system at its optimized nanoindentation parameter condition (R_i = 20 μm and P_{max} = 5 mN). Red dotted curve shows the theoretical prediction of P-h response and the green dotted region infers the region that is dominantly influenced by reversible SIMT in the NiTi alloy. Ptr in the graph highlights the indentation load at which SIMT initiates in the material. (b) Schematic representation of share of different deformation mechanisms within the nanoindentation volume for pseudoelastic NiTi system indented using Ri of 10 μm and 20 μm. (c) $\sigma_{ind} - \varepsilon_{ind}$ curve corresponding to the P-h response (**Figure 7(a)**). Here, σtr is the transformation stress to initiate SIMT [2].*

$$\varepsilon_{ind} \approx \frac{h}{2.4\,a} \quad a = \sqrt{R_i\,h} \tag{10}$$

Prior to conversation of *P-h* results into σ_{ind} - ε_{ind} curve, *ZPC* is performed following Section 6.2.2. **Figure** 7(c) shows the σ_{ind} - ε_{ind} curve that is generated from the *P-h* response of NiTi system. Interestingly the curve has shown the signature trends of pseudoelastic system like sudden changes in the transformation strength, plateau strain, significant recovery etc. The transformation strength (σ_{tr}) of the NiTi system estimated from the nanoindentation resembles reasonably well with that derived from uni-axial compression test [2]. This has validated the present protocol for future analysis on smart characteristics of NiTi based shape memory alloys.

8. Closure

The present chapter elucidates the vast potential of nanoindentation technique to develop insights about the localized stress–strain characteristics of materials. Nevertheless, to achieve the σ_{ind} - ε_{ind} curve, experiments need to be carefully designed. Also, post indention analysis should be meticulously performed to obtain the reliable data. Blunt spherical indenter tip is primarily necessary to activate elastic and plastic mechanisms sequentially in the material and thereby to estimate the σ_{ind} - ε_{ind} curve. On the other hand, different post-indentation analysis has to be adopted based on the mode of nanoindenter and the material of interest to compute the *indentation stress–strain* data. Validation of the protocols are also discussed for pseudoelastic material systems. The detailed explanation provided in the present chapter based on the physical mechanism associated with different alloy system upon indentation and further data analysis can pave the way for future usage of this method of analysis in various studies.

Author details

Indrani Sen[1]* and S. Sujith Kumar[1,2]

1 Department of Metallurgical and Materials Engineering, Indian Institute of Technology, Kharagpur, India

2 Department of Metallurgical Engineering and Materials Science, Indian Institute of Technology, Bombay, India

*Address all correspondence to: indrani.sen@metal.iitkgp.ac.in

References

[1] D. Tabor, The Hardness of Metals, Oxford University Press, 1951.

[2] S. Kumar S, I.A. Kumar, L. Marandi, I. Sen, Assessment of small-scale deformation characteristics and stress-strain behavior of NiTi based shape memory alloy using nanoindentation, Acta Mater. 16375 (2020) 1–2. https://doi.org/10.1016/j.actamat.2020.09.080.

[3] S. Pathak, S.R. Kalidindi, Spherical nanoindentation stress-strain curves, Mater. Sci. Eng. R Reports. 91 (2015) 1–36. https://doi.org/10.1016/j.mser.2015.02.001.

[4] Antony *C. fisher* Cripps, Nanoindentation, Springer International Publishing, 2011.

[5] G.M. Pharr, An improved technique for determining hardness and elastic modulus using load and displacement sensing indentation experiments, J. Mater. Res. 7 (1992) 1564–1583. https://doi.org/10.1557/JMR.1992.1564.

[6] S. Kumar S, L. Marandi, V.K. Balla, S. Bysakh, D. Piorunek, G. Eggeler, M. Das, I. Sen, Microstructure – Property correlations for additively manufactured NiTi based shape memory alloys, Materialia. 8 (2019) 100456. https://doi.org/10.1016/j.mtla.2019.100456.

[7] J. Hu, W. Zhang, G. Peng, T. Zhang, Y. Zhang, Nanoindentation deformation of refine-grained AZ31 magnesium alloy: Indentation size effect, pop-in effect and creep behavior, Mater. Sci. Eng. A. 725 (2018) 522–529. https://doi.org/10.1016/j.msea.2018.03.104.

[8] I.C. Choi, B.G. Yoo, Y.J. Kim, J. Il Jang, Indentation creep revisited, J. Mater. Res. 27 (2012) 3–11. https://doi.org/10.1557/jmr.2011.213.

[9] S. Graça, R. Colaço, P.A. Carvalho, R. Vilar, Determination of dislocation density from hardness measurements in metals, Mater. Lett. 62 (2008) 3812–3814. https://doi.org/10.1016/j.matlet.2008.04.072.

[10] C.P. Frick, T.W. Lang, K. Spark, K. Gall, Stress-induced martensitic transformations and shape memory at nanometer scales, Acta Mater. 54 (2006) 2223–2234. https://doi.org/10.1016/j.actamat.2006.01.030.

[11] S. Sujith Kumar, I. Sen, A Comparative Study on Deformation Behaviour of Superelastic NiTi with Traditional Elastic–Plastic Alloys in Sub-micron Scale, Trans. Indian Inst. Met. (2021). https://doi.org/10.1007/s12666-021-02207-8.

[12] K. Jacob, D. Yadav, S. Dixit, A. Hohenwarter, B.N. Jaya, High pressure torsion processing of maraging steel 250: Microstructure and mechanical behaviour evolution, Mater. Sci. Eng. A. 802 (2021) 140665. https://doi.org/10.1016/j.msea.2020.140665.

[13] W.C. Oliver, G.M. Pharr, Measurement of hardness and elastic modulus by instrumented indentation: Advances in understanding and refinements to methodology, J. Mater. Res. 19 (2004) 3–20. https://doi.org/10.1557/jmr.2004.19.1.3.

[14] S. Pathak, D. Stojakovic, S.R. Kalidindi, Measurement of the local mechanical properties in polycrystalline samples using spherical nanoindentation and orientation imaging microscopy, Acta Mater. 57 (2009) 3020–3028. https://doi.org/10.1016/j.actamat.2009.03.008.

[15] S. Pathak, J. Shaffer, S.R. Kalidindi, Determination of an effective zero-point and extraction of indentation stress-strain curves without the continuous stiffness measurement signal, Scr. Mater. 60 (2009) 439–442.

https://doi.org/10.1016/j.scriptamat.2008.11.028.

[16] S.R. Kalidindi, S. Pathak, Determination of the effective zero-point and the extraction of spherical nanoindentation stress-strain curves, Acta Mater. 56 (2008) 3523–3532. https://doi.org/10.1016/j.actamat.2008.03.036.

[17] N.G. Mathews, A.K. Saxena, C. Kirchlechner, G. Dehm, B.N. Jaya, Effect of size and domain orientation on strength of Barium Titanate, Scr. Mater. 182 (2020) 68–73. https://doi.org/10.1016/j.scriptamat.2020.02.039.

[18] G. Dehm, B.N. Jaya, R. Raghavan, C. Kirchlechner, Overview on micro- and nanomechanical testing: New insights in interface plasticity and fracture at small length scales, Acta Mater. 142 (2018) 248–282. https://doi.org/10.1016/j.actamat.2017.06.019.

[19] J.S. Field, M. V. Swain, A simple predictivity model for spherical indentation, J. Mater. Res. 8 (1993) 297–306. https://doi.org/10.1557/JMR.1993.0297.

[20] S.K. Kang, Y.C. Kim, Y.H. Lee, J.Y. Kim, D. Kwon, Hertz elastic contact in spherical nanoindentation considering infinitesimal deformation of indenter, Tech. Proc. 2012 NSTI Nanotechnol. Conf. Expo, NSTI-Nanotech 2012. 1 (2012) 132–135.

[21] B.C. Maji, M. Krishnan, The effect of microstructure on the shape recovery of a Fe-Mn-Si-Cr-Ni stainless steel shape memory alloy, Scr. Mater. 48 (2003) 71–77. https://doi.org/10.1016/S1359-6462(02)00348-2.

[22] X. Li, B. Bhushan, A review of nanoindentation continuous stiffness measurement technique and its applications, Mater. Charact. 48 (2002) 11–36. https://doi.org/10.1016/S1044-5803(02)00192-4.

Chapter 10

Toward an Instrumented Strength Microprobe – Origins of the Oliver-Pharr Method and Continued Advancements in Nanoindentation: Part 1

Bryer C. Sousa, Jennifer Hay and Danielle L. Cote

Abstract

Sub-micron instrumented indentation testing and standardized nanoindentation testing systems have become commonplace within the materials engineering community. Though commonly utilized for mechanical characterization, general appreciation and understanding of the governing theory, formulations and best practices underpinning modern nanoindentation systems appears to remain relatively elusive to the general materials science and engineering community as well as nanoindentation practitioners using such systems for mechanical assessment. Accordingly, the present chapter details how nanoindentation methods emerged and how the Oliver-Pharr method of nanoindentation testing and analysis was constructed and refined to yield theoretically consistent and readily implementable attributes for probing small-scale mechanical properties via microscopy free indentation testing.

Keywords: nanoindentation, depth-sensing indentation testing, instrumented indentation testing, Oliver-Pharr method, hardness, modulus, load-displacement

1. Introduction

Motivated by the need for a consistent and implementable method for performing sub-micron instrumented indentation tests (IIT) on materials, as well as for analyzing IIT data at the micro- and nanometer scales, Oliver and Pharr (O&P) improved upon the work of Doerner and Nix (D&N) and others in Ref. [1]. They focused on the work of D&N [2] because they found that D&N's assumption of linearity in the upper one-third of the indentation load-displacement curve was inconsistent with O&P's observations across various materials subject to small-scale IIT. Prior to the publication of O&P's findings in 1992, the Doerner-Nix (DN) method was considered the most comprehensive approach for determining hardness (H) and elastic modulus (E), but it was replaced by the Oliver-Pharr (OP) method after publication.

Motivated by the need for a theoretically consistent and practically implementable modality of performing sub-micron or small volume instrumented indentation testing (IIT) of materials, as well as the need for a method of analyzing IIT test data at the micrometer and nanometer length scales, Oliver and Pharr (O&P) refined and considered the foundational work of Doerner and Nix (D&N), among others, in Ref. [1]. O&P gave particular attention to the work of D&N [2] due to O&P's discovery that D&N's assumed linearity of the upper one-third of a given indentation load-displacement unloading curve was at odds with O&P's regular experimental observations across numerable material types. Until O&P published their findings in 1992, the Doerner-Nix (DN) method for determining both hardness (*H*) and modulus of elasticity (*E*), or Young's modulus, was thought to be the most thorough load-displacement data analysis approach prior to the introduction of the Oliver-Pharr (OP) method of testing and analysis.

With such significance and implications in mind, O&P set out to address standing issues, problems, and inadequacies associated with the DN method of sub-micron indentation testing and data analysis. Beyond simply demonstrating that unloading curves are rarely, if at all, linear, O&P went on to substantiate their hypothesis that unloading curves ought to be thought of as nonlinear and power-law-based. O&P presented load-displacement data for an array of materials (ranging from crystalline ceramics to amorphous glasses as well as both soft and hard cubic-centered metals) to demonstrate said non-linearity of unloading data [1].

In addition to the presentation of such substantiating nanoindentation load (*P*) vs. displacement or depth (*h*) data, data analysis, and resultant findings, the OP method was carefully documented such that physically justifiable indentation depth determination was reliably and repeatably procurable. Furthermore, the resultant abilities brought about by the OP method were further detailed for subsequent use in establishing peak applied load contact areas and contact area functions for various indenter tips and tip geometries. The OP method also provided beneficiaries of nanoindentation testing with a measurement and analysis heuristic for depth determination with load-controlled nanoindentation testing and load-controlled nanoindentation-derived data analysis. Remarkably, O&P had done so while avoiding the use of (or need for) post-indentation microscopy, which not only remains both time-consuming and costly for nanometric resolutions but was also generally out of the reach for many researchers and engineers during the early 1990s.

Upon rendering such findings, *H* and *E* values were then deduced via load-displacement data, as analyzed according to said OP methodology, and then compared with values derived from alternative and independent means to demonstrate the accuracy of the OP method [1]. A discussion was also presented in 1992, which pragmatically coupled theory and practice together, such that load frame compliance and indenter shape functions could be integrated into nanoindentation load-displacement data analysis platforms and experimental nanoindentation frameworks generally.

2. Consideration of the Oliver-Pharr approach

In analytical terms, *h* represents the total nanoindenter displacement, defined mathematically in Eq. (1), such that

$$h = h_c + h_s \tag{1}$$

wherein h_c is the contact depth, or the distance under which indenter tip contact is made normal to the sample surface, and h_s represents the surface displacement (which is now classified in the literature as pile-up, sink-in, or both, depending upon the material deformation mechanics, among other factors) about the contact perimeter. In addition to h, h_c, and h_s, P_{max} captures the load applied during nanoindentation testing at h_{max}, or the maximal IIT displacement achieved during a given test. At the same time, O&P concurrently considered S, a, and h_f too, such that S_{max} was defined as the experimentally measured stiffness (S) obtained via the slope of the tangent line procured from the initial unloading point along the curve, which occurs upon reaching P_{max} at h_{max}.

That said, h_f was defined as the depth of the residual IIT or nanoindentation impression upon both complete unloading and total indenter removal from the specimen, whereas a was presented as a surrogate geometrical contact radius. Considering the discussion presented thus far, **Figure 1(a)** presents load-displacement data at various points along loading and unloading curves. In addition, **Figure 1(a)** also captures their relation to the inelastic work and the work of elastic deformation associated with a given indentation test. Finally, **Figure 1(b)** also presents a cross-sectional view of indentation phenomena.

In addition to Eq. (1), the OP method determines E as expressed in Eq. (2), such that

$$\frac{1}{E_r} = \frac{1 - v^2}{E} + \frac{1 - v^2}{E_i} \tag{2}$$

wherein E_r is the reduced modulus, ν is the Poisson's ratio of the specimen, ν_i is the Poisson's ratio of the indenter tip, and E_i is the modulus of elasticity associated with the indenter tip material too. Since v_i and E_i are known apriori (let us assume diamond as Eq. (2) is recast for simplicity and consistency with the conventional material utilized as nanoindenter tips), E_r can be solved via another relation and after that substituted back into Eq. (2) such that E can be obtained through

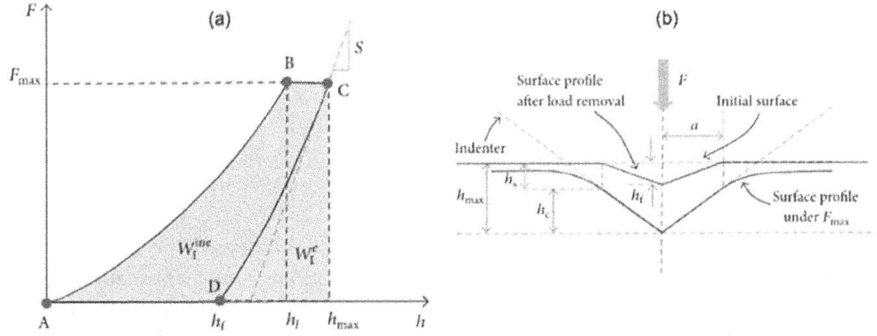

Figure 1.
In (a), A is the initial contact of the indenter tip with the tested material; point B is the point at which the P_{max} is reached; C is the point at which unloading from P_{max} begins after P_{max} is held for a predefined time to account for the influence of creep; D captures h_f, which is reached upon complete unloading; W_I^{ine} is the inelastic work of indentation; W_I^e is the elastic work of indentation; h_l represents the depth associated with B ($h_l = h_{max}$ when creep is not accounted for); S is the experimentally measured stiffness obtained via the slope of the tangent line procured from the load-displacement nanoindentation curve. (b) Presents a cross-sectional view of indentation testing related phenomena (loading and unloading, etc.). Both (a) and (b) were sourced from Ref. [3].

algebraic manipulation and arithmetic. For commonly used diamond indenter tips, we may re-render Eq. (2) to yield Eq. (3):

$$\frac{1}{E_r} = \frac{1-v^2}{E} + \frac{1-v^2}{E_i}$$
$$\Longrightarrow \frac{1-v^2}{E} = \frac{1}{E_r} - \frac{1-v_i^2}{E_i}$$
$$\therefore E = \left(1-v^2\right)\left(\frac{1}{E_r} - \frac{1-v_i^2}{E_i}\right)^{-1}$$
$$\therefore E = \left(1-v^2\right)\left(\frac{1}{E_r} - \frac{1-(0.07)^2}{1140\text{ GPa}}\right)^{-1}$$
$$\therefore E = \left(1-v^2\right)\left(\frac{1}{E_r} - \frac{1-0.9951}{1140\text{ GPa}}\right)^{-1} \tag{3}$$

Thus, with knowledge of the test specimen's v and an empirical means of identifying the E_r from load-depth data collection and analysis, E may be readily attained for a given material. That said, Bulychev et al. [4], and others, according to Poon et al. [5], recognized the following relationship between P, h, S, E_r and the projected area of elastic contact (which had been refined and eventually referenced as the contact area per the OP method), expressed as A in Eq. (4), such that

$$S = \frac{dP}{dh}$$
$$= \frac{2E_r\sqrt{A}}{\sqrt{\pi}} \tag{4}$$

Consequently, solving for E_r yields Eq. (5), wherein

$$E_r = \frac{S\sqrt{\pi}}{2\sqrt{A}} \tag{5}$$

and Eq. (5) may therefore be substituted into the previously documented expression for E, which was given in Eq. (2), such that Eq. (6) yields

$$E = 1 - \frac{v^2}{\left(\frac{S\sqrt{\pi}}{2\sqrt{A}}\right)^{-1} - \frac{1-v_i^2}{E_i}} \tag{6}$$

The mathematical manipulation of said equivalence relations serve as a means of enabling O&P's pedogeological improvements, methodology, and approach to be more explicitly considered. Returning to the matter of fitting loading or unloading segments of load-displacement or load-depth curves via raw nanoindentation data, note that O&P extensively relied upon, as well as critically considered, Sneddon's relation between load and depth for basic punch geometries [6], as presented herein through Eq. (7), wherein

$$P = \alpha h^m \tag{7}$$

In Eq. (7), α, as well as m, are both constants whose values depend on tip geometry. In the case of a flat cylindrical punch geometry, $m = 1.0$. However, in the case of a conical punch geometry, $m = 2.0$. Additionally, $m = 1.5$ for both spheres at shallow indentation depths and paraboloids of revolution too. One ought to note that Eq. (7) is explicitly related to the loading portion of the complete loading and unloading nanoindentation cycle. Beyond simply revisiting the matter of loading curves, one may also note that the mathematical expression H, which was passively noted above as a discernible mechanical performance indicator (rather than an actual material property [7]) for a given material, may be rendered as Eq. (8), wherein

$$H = \frac{P_{h_x}}{A} \tag{8}$$

Importantly, O&P also assumed that a given indenter tip geometry might be expressed in terms of an area function, $F(h)$, that relates h with the cross-sectional area of a respective tip. O&P continued their presentation of $F(h)$ through the lens of h_c, given the practical importance of h_c in nanoindentation testing and data analysis, rather than area relations associated with the distance away from the apex of a probe, or h, such that Eq. (9) may be formulated as

$$A = F(h_c) \tag{9}$$

It is also important to note that O&P regularly reiterated the mathematically expressed rendition of $F(h_c)$, given in Eq. (9), must be determined using the contact area function calibration approach detailed in Ref. [1] before pursuing the application (or use) of the OP data analysis method. Thus, recall that in Eq. (9), $F(h_c)$ functionally relied upon h_c as a means of assessing A. Hence, one ought to observe the fact that h_c can be expressed as a function of h_{max} and h_s, such that Eq. (10) can be presented as

$$h_c = h_{max} - h_s \tag{10}$$

and thus, pairs well with Eq. (1) too. Nevertheless, the means of ascertaining h_s was noted as being remarkably dependent upon the exact nanoindenter tip geometry utilized. Assuming conical geometries, O&P invoked another expression presented by Sneddon in the 1960s, such that Eq. (11) yields

$$h_s = \frac{\pi - 2}{\pi}\left(h - h_f\right) \tag{11}$$

By way of once again considering one of O&P's intellectual precursors, Ian N. Sneddon of the Department of Mathematics of the University of Glasgow in Scotland, O&P remedied the use of the quantity $\left(h - h_f\right)$ in Eq. (11) by way of also invoking Sneddon's load-depth equivalence relation for conical tip geometries, which may be given as Eq. (12), wherein

$$\left(h - h_f\right) = \frac{2P}{S} \tag{12}$$

While it may not seem readily evident as to why Eq. (11) must have been presented in terms of the quantity $\left(h - h_f\right)$, rather than h alone, recall the fact that Sneddon's solutions only hold for elastic indentation displacement, rather than the total (i.e., both plastic and elastic portions of indentation deformation phenomena) displacement, which are of course convoluted with one another in unprocessed and experimentally recorded displacement data from a nanoindenter. Thus, the clever use of Eq. (12) enabled substitutional elimination of the respective $(h - h_f)$ quantity altogether. As a result of the substitution of Eq. (12) into Eq. (11), Eq. (13) can be expressed as follows

$$\begin{aligned} h_s &= \frac{\pi - 2}{\pi}(h - h_f) \\ &= \frac{\pi - 2}{\pi}\left(\frac{2P}{S}\right) \\ &= \frac{\epsilon P}{S} \end{aligned} \tag{13}$$

Therefore, when evaluated at P_{max}, Eq. (13) may be rewritten as Eq. (14), where

$$h_s|_{P=P_{max}} = \epsilon \frac{P_{max}}{S} \tag{14}$$

Though ϵ was substituted into Eq. (13) above in place of $2\pi^{-1}(\pi - 2)$, ϵ is variable across tip geometries in a comparable manner to that of m in the case of Eq. (7). Thus, one ought to note that in the case of a flat cylindrical punch geometry, $\epsilon = 1$. However, in the case of a conical punch geometry, $\epsilon = 0.72$. Additionally, $\epsilon = 0.75$ for both spheres at shallow indentation depths and paraboloids of revolution. Also noted earlier was the fact that O&P harped upon the misappropriated assumption of initial unloading load-depth curve linearity by D&N. Alongside such remarks concerned with said critical commentary by O&P, the pair of researchers put forth their simple power-law unloading relation in 1992, as shown in Eq. (15) herein, such that

$$P = Q(h - h_f)^k \tag{15}$$

wherein all but P and h are determined via applying a least-squares fitting procedure. For comparison with the linear fitting procedure associated with the precursory D&N method of S quantification from sub-micron IIT, O&P presented an underappreciated graphical figure plotted within [1], wherein O&P presented the unloading stiffness obtained for a tungsten specimen as a function of the fraction of the unloading curve considered during data analysis and as a function of using either the DN or OP stiffness determination method applied. Said graphical rendering of peak load stiffness values determined using linear fitting approaches compared to O&P's power law fitting method, coupled with O&P's presentation of the constant stiffness values as a function of the unloading curve considered during data fitting attests to the methodological enhancements and integrity underpinning data interpretation by O&P [1].

In any case, a load frame compliance determination procedure is the next aspect of the OP method to be contextualized and considered herein. For those unfamiliar with the peculiarities of nanoindentation (and load-controlled nanoindenter systems as well as their assembly as a class of scientific instrumentation), the load frame compliance, let alone the importance of such a value, stems from the instrumented method, or modality, incorporated for displacement sensing capabilities via load-controlled systems. More to the point, when an automated and OP compatible IIT characterization suite or nanoindenter is utilized, the raw displacement information recorded by the system must be considered and corrected for both the load frame and the specimen being tested. In turn, precise knowledge of the load frame compliance term enables the unprocessed displacement data recorded by the system controller to be deconvoluted, algebraically speaking (rather than statistically speaking), such that the displacement contribution from specimen deformation via indentation loading and the displacement contribution from the load frame can be isolated from one another. Consequently, specimen displacements eventually yield indentation depths associated with their respectively applied loads.

However, the mathematics underpinning said displacement deconvolution and analysis, or displacement contribution decoupling, begins with two basic assumptions. The first assumption is that E values are depth-independent, whereas the second assumption is that one may treat said dual compliance terms as a pair of springs in a series, such that Eq. (16) yields

$$C = C_s + C_f \tag{16}$$

wherein C is the measured compliance, C_s is the specimen compliance and C_f is the load frame compliance term. Within the realm of contact mechanics and a given materials' elastic and plastic deformation mechanisms, contact stiffness may be expressed as the inverse of compliance, and compliance may therefore be expressed as the inverse of stiffness. By way of such definitions, in conjunction with Eq. (4) and the spring series-inspired and assumed relation, the substitution of S^{-1} in place of C_s yields Eq. (17), wherein

$$\begin{aligned} S &= \frac{2E_r\sqrt{A}}{\sqrt{\pi}} \\ &\rightarrow C_s = S^{-1} \\ S^{-1} &= \frac{1}{S} \\ &= \frac{\sqrt{\pi}}{2E_r\sqrt{A}} \\ \therefore C &= C_s + C_f \\ C &= \frac{\sqrt{\pi}}{2E_r\sqrt{A}} + C_f \end{aligned} \tag{17}$$

By invoking the additional assumption explicitly stated earlier surrounding E's insensitivity to depth, one may proportionally reformulate Eq. (17) as follows Eq. (18):

$$\begin{aligned} C &= C_s + C_f \\ &= \frac{\sqrt{\pi}}{2E_r\sqrt{A}} + C_f \\ \therefore C &= \frac{\sqrt{\pi}}{2E_r\sqrt{A}} + C_f \\ C &= \frac{\sqrt{\pi}}{2E_r}\frac{1}{\sqrt{A}} + C_f \\ &= x\frac{1}{\sqrt{A}} + C_f \\ &= \frac{x}{\sqrt{A}} + C_f \end{aligned}$$

$$C \propto \left(\sqrt{A}\right)^{-1} + C_f$$

$$\therefore C \propto \frac{1}{\sqrt{A}} + C_f \tag{18}$$

As a result, empirically fitting C as a function of $A^{-0.5}$ to a linear expression yields an intercept equal to the load frame compliance contribution. After that, proportional relations may once again be revisited, alongside aforementioned equivalence relations, such that Eq. (18) yields Eq. (19), such that

$$\begin{aligned} C &\propto \frac{1}{\sqrt{A}} + C_f \\ &\propto \frac{1}{\sqrt{A}} \\ &\propto \frac{1}{\sqrt{F(h_c)}} \\ &\propto \frac{1}{\sqrt{24.5h_c^2}} \end{aligned} \tag{19}$$

As such, Eq. (19) captures an idealized Berkovich indenter tip geometry (that is, the widely used, three-sided pyramidal tip geometry) such that the curve-fitting procedure associated with C as a function of $A^{-0.5}$ may be pursued for estimating C_f and E_r given the assumption and use of an idealized and perfect Berkovich indenter tip's contact area function. Since C_f and E_r were assumed to be constant as a function of C_s, n-number of indentation size-specific and geometrically ideal contact area values were calculated by the O&P procedure and are easily calculated by modern systems today, via rearrangement of the already presented equations, as shown below via Eq. (20):

$$\begin{aligned} C &= C_s + C_f \\ &= \frac{\sqrt{\pi}}{2E_r\sqrt{A}} + C_f \\ \rightarrow (C - C_f) &= \frac{\sqrt{\pi}}{2E_r\sqrt{A}} \\ \rightarrow \sqrt{A}(C - C_f) &= \frac{\sqrt{\pi}}{2E_r} \\ \therefore \sqrt{A} &= \frac{\sqrt{\pi}}{2E_r}(C - C_f)^{-1} \\ \rightarrow \sqrt{A} &= \frac{\sqrt{\pi}}{2E_r(C - C_f)} \\ \therefore \left(\sqrt{A} \right.&= \left.\frac{\sqrt{\pi}}{2E_r(C - C_f)}\right)^2 \\ \rightarrow A &= \frac{\pi}{4E_r^2(C - C_f)^2} \end{aligned}$$

$$\begin{aligned}
&\therefore C \propto \frac{1}{\sqrt{A}} + C_f \\
&\Rightarrow A = F(h_c) \\
&= x^2 (C - C_f)^{-2} \\
&\rightarrow F(h_c) = x^2 \frac{1}{(C_s(h_c))^2} \\
&= \frac{\pi}{4E_r^2 (C_s(h_c))^2}
\end{aligned} \tag{20}$$

Accordingly, when the proportionality just presented between the total compliance, as a function of indentation contact depth, and the contact area function as a function of indentation contact, is coupled with surrogate specimen data as well as analytical fitting procedures for the estimation of a calibrated contact area function, machine, tip, and reference specimen dependent calibrations may be pursued on an as-needed basis by a given researcher. In any case, the contact area function analytical fitting procedure may be conceptualized as an up to eight-parameter harmonic average of polynomials, which is expressed as Eq. (21), such that

$$F(h_c) = (24.5)h_c^2 + C_1 h_c + C_2 h_c^{\frac{1}{2}} + C_3 h_c^{\frac{1}{4}} + \cdots + C_8 h_c^{\frac{1}{128}} \tag{21}$$

More to the point, the eight-parameter harmonic average of polynomials given and expressed above takes the mathematical form of a series expression that can be given as Eq. (22) as follows:

$$F(h_c) = \sum_{n=0}^{8} C_n (h_c)^{2-n} \tag{22}$$

The mathematical formulations pursued herein comprise the various aspects of the 1992 method presented by O&P, since $A = F(h_c)$ enables mechanical properties of materials, i.e., E and H, to be assessed at nanomechanical and micromechanical length scales without the need for microscopy-based evaluation of contact areas on a per measurement basis.

Also, worth consideration herein is the appendix to O&P's 1992 manuscript, which was titled Continuous Measurement of Contact Stiffness by a Dynamic Technique. Though dynamic/CSM nanoindentation has emerged as a valuable tool throughout the modern instrumented indentation and nanoindentation community [8], the limited realization of O&P's CSM technique, relative to the time of original publication, reveals scattered attempts by contemporaries in the mid-to-late 1990s to wrestle with not only the OP method and associated manuscript but also the intellectual novelty underpinning the appended framework, as well as the mechanical properties that were eventually found to be measurable via the incorporation of the details presented in the 1992 appendix on dynamic nanoindentation too. That said, forthcoming areas of the present chapter provide details surrounding the wide range of mechanical properties that became experimentally probable due to CSM-based nanoindentation methods. In the meantime, the needs addressed by O&P's peer-reviewed research deliverables detailed in their 1992 manuscript were considered next.

3. Needs addressed by the Oliver-Pharr method

As noted by O&P in 1992, the emergence of thin-film technologies as well as technological advancements, which directly followed from functional thin-film material development and advancements too, ultimately inspired recognition of the fact that sensing load-displacement behavior via indentation enabled thin film mechanical properties and layered material property extraction [1]. Since such a statement from O&P was concerned with need-based sources of inspiration associated with the early 1980s, i.e., a decade or so before the publication of concern, work undertaken before the present 1992 article attempted to address such needs as effectively as possible.

Needs remained through 1992 and were therefore accordingly addressed, in part, by O&P. However, said needs were separable (categorically speaking) into industrial or materials characterization-based needs and pedagogical or methodological needs. Stated otherwise, needs were classifiable by either the application of nanoindentation, such as micron-scale IIT, or the development and research into nanoindentation as a sub-field of explicit study. Of course, it stands to reason that those advancements in our understanding of nanoindentation, without an underpinning industrial driving force, would nevertheless enable further materials development and research to thrive too. However, without industrial interest and backing, it may very well have taken much more time to reach its' industrially minded status and capabilities when viewed through the lens of technology readiness levels.

Still, the work of O&P was further motivated by materials engineering and science needs within the broader governmental, industrial, academic, collaborative, and technological ecosystem. That being said, one point that ought to be emphasized before further consideration of the emergence of OP mode of nanoindentation testing and analysis is the fact that regardless of the applied materials characterization needs (that is, thin films versus nuclear materials, for example), nanoindentation remains sought after to date, since IIT analysis at nanometric length scales enables small-scale mechanical properties to be explored in a systematic and reproducible manner (so long as the OP method is adhered to).

Consequently, one may reason that, as defined in O&P's original research article, nanoindentation was formalized and developed in commercial forms to serve as a foundational linkage between small-scale properties and mesoscale material behavior. Stated another way, nanoindentation and the OP method of load-displacement data analysis served to, at least in part, address the standing need for probing nanomechanical and micromechanical material strength and mechanical properties alike, regardless of a particular material's industrial intricacies. Of course, industrial sectors beyond applied monolithic materials-based component processing and production would ultimately be influenced by nanoindentation testing too. Such influences were found to be true given nanoindentation's widespread and successful adoption by the polymeric [9], biological [10], and composite material sectors, too [11].

In keeping with the passage above, recall once more that O&P attempted to address the need for a small-scale mechanical property evaluation method and an experimental system to do so, which became known as a nanoindenter. As stated by O&P, such an instrument could be classified as a mechanical property's microprobe [1]. Indeed, small-scale capabilities were reportedly achieved by O&P in 1992, wherein the Nano Instruments, Inc., Nanoindenter, which was housed at Oak Ridge National Laboratory (Oak Ridge, TN, USA), achieved a load resolution of 0.3 μN and a displacement resolution of 0.16 nm; in turn, maintaining sub-micrometer and therefore small-scale resolution.

Having presented the mathematical analysis and manipulation underlying O&P's 1992 approach to nanoindentation data analysis, in addition to **Figure 1(b)**'s rendering of a cross-section of an indenter tip loaded upon a hypothetical target material or specimen, note that additional visualizations or schematics are provided in Ref. [1] too. O&P presented a schematic graphical representation of a load vs. depth nanoindentation curve for further context; a schematic detailing how the Nano Instruments, Inc., Nanoindenter was assembled; and an SEM image or micrograph of nanoindented fused Si specimen after reaching a maximal applied load of 40 mN [1], for instance.

O&P also stated that practical reasons, such as cost and difficulty of resolving micrographs of shallow indents at the time, underpinned the need for calculating the geometry of indentation-induced residual impressions such that contact areas could be determined [1]. Such a statement attests to another need-based thought process by O&P, wherein they document how effectively microscopy-free nanoindentation was able to be procured by O&P in 1992. Reflecting on such a need in 2004, O&P also noted the fact that their 1992 methodology for hardness and modulus determination from load-depth data via ITT inspired techniques had been widely adopted for small-scale mechanical characterization of materials, suggesting that said widespread adoption was largely due to the microscopy-free nature of the OP method [12].

Returning to materialistic needs addressed by O&P, Chen et al. noted that mechanical property determination of thin films was mainly driven by the needs and desires expressed semiconductor and magnetic storage materials engineers [13]. Prior to the work of Chen et al., Menčík et al. noted that a significant mechanical property of interest to the thin film community was the hardness of the thin film and the thin film elastic modulus [14]. The determination of the modulus of elasticity associated with a functional thin film was critical to the evaluation of residual stresses via X-ray based analysis; the deduction of deformation-driven thermomechanical stress accumulation within the thin film as the film-substrate component is subject to an externally applied load; and even the determination of delamination mechanics [14].

Thus, one need not be surprised that even though O&P only made mention of thin-film mechanics in 1992, to identify motivations for nanoindentation research and development, the OP method of analysis emerged as a standard nanoindentation load-displacement data analysis framework that became [15], and remains [16], commonly applied to thin films. Nevertheless, as noted by Saha and Nix a decade after O&P's 1992 article, O&P formulated the OP method using monolithic materials, i.e., non-composite-like components [17]. Regardless, one may still reason that O&P indirectly, although intentionally, laid the experimental groundwork for thin-film mechanical property evaluation via IIT modes of analysis, which has since been refined further following 1992 innumerable works, such as those associated with [18]. Additionally, one may also note the fact that the need for sub-micrometric mechanical characterization also followed from the limitations of the otherwise employed micro-beam bending and film deflection testing methods for thin-film elastic modulus assessment [19], among other approaches, which were identified in 1990 by Alexopoulos et al. [20].

4. State of the art prior to Oliver-Pharr methodology

Before the publication of O&P's 1992 research article, which has since matured into one of the most highly cited and influential manuscripts in the field of materials science and engineering to date, nanoindentation as a commercialized technological advent was developed by John Pethica, Ron Hutchings, and Warren Oliver in 1983 while Pethica, Hutchings and Oliver were working together at Brown Boveri in Switzerland [21]. Accordingly, such a timeline provides a relatively lower bound for prior work considerations in the decade preceding the 1992 publication alongside the understanding of nanoindentation garnered after Pethica et al. initiated their collaboration in the 1980s [22]. Hence, the present chapter refocuses upon the state of the art between the innovative development of a nanoindenter/IIT in the early 1980s and O&P's 1992 article. The remaining portion of the present section is dedicated to nanoindentation or IIT developments between 1982 and 1992.

In so far as work related to the state-of-the-art surrounding indentation analysis before OP methodological formalization, prior work may be initiated herein via considering the work of Newey et al., which was published in 1982 [23]. Specifically, Newey et al. documented an ultra-low-load penetration hardness tester and testing approach that also employed a non-optical method of residual indent geometry deduction while continuously recording indenter tip penetration depths well as applied loads too. Remarkably, Newey et al.'s ultramicrohardness tester achieved a load resolution of 10 μN, a depth resolution of approximately 5 nm, and a maximal load of 3 mN. Moreover, unlike Newey et al.'s counterparts, such as Nishibori et al. in Refs. [24–26], or Frohlich et al. [27], Newey et al.'s approach resulted in indenter probe penetration depth recording as a function of the load applied and the time required to reach said load, therefore enabling time- and load-dependent indentation-induced mechanisms to be observed.

At the same time, Newey et al.'s ultra-low-load penetration hardness tester was able to capture an indicator of material elasticity and even adhesion properties via testing too. Thus, like O&P's objectives surrounding the development of an understanding as to how to formulate more wear-resistant metallic surfaces via ion implantation for tribological enhancement [21], Newey et al.'s advancements were to investigate hardness as a function of ion-based implantation processing of materials.

Newey et al.'s approach agreed with the precedent established by E. S. Berkovich surrounding the suitability of three-sided pyramidal tip geometries over that of Knoop and Vickers tip geometries because of the inherent fact that Berkovich tips (i.e., a particular form of three-sided pyramidal tips) meet at only one apex point. Moreover, Newey et al. invoked the proportionality shared between pyramidal indentation depth and applied load when an (assumed) ideal plastic material was undergoing indentation testing, such that the following theoretical relation (Eq. (23)) between hardness (H_v), force (F), and depth (δ), in base units, was turned to, such that

$$H_v = \frac{0.0378(F)}{\delta^2} \tag{23}$$

Later, Eq. (23) was amended to include on-load and off-load hardness analysis by way of including elastic and plastic indentation contributions via the use of $\delta_T - \delta_e$ in place of δ; thus, overcoming the assumption of an idealized, fully plastic material, which had been reflected in Eq. (23). In all, Newey et al. noted that the off-load depth of $\delta_T - \delta_e$ maintained a 5–10% difference with an independently assessed depth, denoted as δ_A, and obtained via ex-situ microscopy analysis post-indentation. However, one must note that the work of Newey et al. was limited to indium (primarily) and electropolished AISI 52100 steel and rock salt, thus prohibiting the 5–10% difference acquired be fully generalize-able across metallic, ceramic, composite materials too.

Moreover, the 5–10% difference between $\delta_T - \delta_e$ in contrast with δ_A, when substituted into Eq. (23) in place of δ, resulted in an overestimation of the hardness by 10–20% when their non-microscopy or microscopy-free depth determination method was invoked. Still, Newey et al. quickly addressed such a discrepancy between their microscopy-free depth determination and microscopy-based counterbalance through the lens of pile-up effects; therefore, suggesting that the 10–20% difference in H_v (i.e., the Vickers hardness number or value) as a function of δ was an artifact of overestimated microscopy-based depths. Finally, Newey et al. correctly pointed out that a phenomenon concerning the non-transferable nature of hardness values obtained using their ultra-microhardness tester, in contrast with a Vickers or Knoop indenter, could be explained through the lens of mean contact pressure or indentation hardness value depth dependence. In doing so, Newey et al. stumbled upon what would later be articulated as an indentation size effect (ISE) that came to be described through the lens of strain gradient plasticity and ultimately unveiled subsequent work by Nix and Gao [28].

Shortly after Newey et al.'s 1982 article, Pethicai et al. highlighted an indentation advancement in a 1983 article focused upon the realm of hardness testing at depths as low as 20 nm [22]. In doing so, Pethicai et al. utilized Ni, Au, and Si to demonstrate that indentation contact areas determined via post-indentation electron microscopy were quantifiable when coupled with the 1908-rooted mathematical formulation of Meyer's hardness. Interestingly, just as Newey et al. noted an ISE prior to Nix and Gao's strain gradient plasticity framework, Pethicai et al. also noted an increase of IIT hardness as a function of smaller indentation size at the submicron length scale [22]. In addition, Pethicai et al. observed relatively increased hardness values at shallow indentation depths and relatively decreased hardness values at greater indentation depths. Finally, like the ultramicrohardness tester developed by Newey et al., indentation load-depth relations were also continuously recorded by Pethicai et al. via loading and unloading cycles such that a quantitative understanding of elastic relaxation could be formulated.

Building upon Pethicai et al.'s novel advancement of indentation abilities into the nm depth regime, Oliver published a subsequent article in 1986, which noted that the technological groundwork, hardware, and understanding of the indentation process had been under development [29]. Said statement by Oliver highlights the pre-1992 state of the art surrounding small volume mechanical property inspection by the mid-1980s. The limited degree of understanding within the respective research community highlights the significant gap addressed by O&P in their 1992 manuscript. Oliver noted that indent geometry determination was not only particularly difficult through microscopy but was also connected to the most critical parameter in relation to contact area determination relative to the specimen and indenter tip. Oliver went so far as to state that a mechanical properties microprobe was not only conceptually exciting but was also being complemented by concomitant advancements in understanding and hardware needed to actualize a submicron resolution system for recording mechanical response and behavior during indentation. Such an assertion readily situates the implications of Oliver's ongoing efforts at the time.

In keeping with the trend of developing practical submicron indentation testing, Doerner et al. considered thin-film plasticity properties compared to substrate curvature techniques [30]. Around the time that Doerner et al. published their submicron indentation testing of thin films and respective findings, Doerner et al. also proposed a nanoindentation data analysis and interpretation framework that was purportedly based upon the use of the commercial nanoindentation system from Nano Instruments, Inc. [2]. Continuation along similar lines to the work detailed in Ref. [30] was also pursued by Oliver et al. in 1987 [31]. During that very same year, the interactive forces associated with a microprobe or nanoprobe indenter tip and a target specimen with a flat surface, as well as the tip-specimen surface responses, were documented in Ref. [32]. Consequently, Pethica and Oliver demonstrated that a true contact area was discernible when local surface stiffness values were measured via the application of an alternating current force to the indenter tip. Said findings were explored beyond nanoindentation and the concurrent emergence of atomic force microscopes and scanning tunneling microscopy enclosed tips.

Consideration of state of the art, pre-1992, must also include the eventual patent and patent-related documentation associated with the commercialization of the Nanoindenter by Pethica and Oliver as of 1989, which was entitled as follows in Ref. [33]: A Method for Continuous Determination of the Elastic Stiffness of Contact Between two Bodies. Beyond the description afforded from the title provided, Pethica and Oliver stated that an ultra-low load IIT/nanoindentation system known as the Nanoindenter and commercially available through Microsciences, Inc. (Norwell, MA, USA) was substantially modified as part of the invention detailed as part of the patent. Said modification to the Nanoindenter allowed the force to be linearly modulated (increased or decreased) via electromotive means throughout various loading rates. The further modification enabled a capacitive displacement gage to be used to determine the indentation area as a function of indenter displacement following initial contact between a target specimen surface and an indenter probe. With said modifications in mind and others detailed directly within the patent under consideration, Pethica and Oliver ultimately devised a means of continuously determining the elastic contact stiffness between two bodies.

Around the point in time that the patent was assigned to the inventors and the U.S. Department of Energy, the materials science and engineering research community poised to benefit from nanoindentation centered characterization capabilities began applying initial approaches to load-displacement data analysis, which concomitantly started to emerge alongside the protocols laid out by O&P in the early 1990s. Furthermore, said research community started coupling the utilization of preliminary, or initialized, approaches to load-displacement data analysis with early versions of the commercially available Nanoindenter and low-load indenters generally; ultimately, applying them to various material systems. For example, Stone et al. published findings surrounding their application of such a mechanical properties microprobe as detailed in Ref. [34]. Specifically, Stone et al. applied continuous indentation testing to sputter-deposited Al thin films adhered to Si substrates.

In addition to the work of Stone et al., Loubet et al. built their micro-indenter-that could record load-depth curves, including both loading and unloading load-displacement curves, to explore the complex phenomena underlying MgO Vickers indentation data in Ref. [35]. Another example was detailed in Ref. [36], wherein Pharr and Oliver applied the state-of-the-art understanding and their own methodological improvements to IIT data analysis and testing pre-1992 to directly link hardness as a function of depth with dislocation structures in a single crystal Ag specimen. Such exploration was performed to contextualize better deformation mechanisms, elasto-plasticity, and plasticity in a pure metal specimen.

Ultimately, during the final years preceding the publication of O&P's paper in 1992, state of the art surrounding small-scale IIT was primarily found to be concerned with refining and proposing physical, computational, foundational, empirical, numerical, and/or theoretical relationships to yield a mechanistic abstraction for experimentally consistent models, which could be used in IIT data analysis. Nevertheless, the clear need for the OP nanoindentation testing methodology and load-depth data analysis protocol can be adequately appreciated through the simple fact that their novel approach avoided any superfluous explanations and instead focused on analytical patterns that could be discerned and replicated by others time and again, as will be discussed in Part 2.

5. Developments immediately following Oliver-Pharr's method

Between 1992 and 2002, there were numerable application-specific and application-inspired uses of nanoindentation and OP-based method and analysis use cases that could have been widely considered herein [37]. However, in so far as the post-OP publication state-of-the-art may be considered, one ought to note that enumerable investigations went on to critically examine the 1992 article by O&P and the OP testing and analysis method or framework. In conjunction with others during the respective period, some critical examinations formulated novel physical and mechanical relationships to extend the range of possibilities associated with sub-micron indentation deformation.

Therefore, one may begin the consideration of the respective decade following O&P's 1992 article by invoking the 1993 article, entitled Mechanical Characterization Using Indentation Experiments, by Oliver et al. [38]. Remarkably, by 1993, Oliver et al. noted that nanoindentation-based methods for assessing the creep stress exponent were formulated; therefore, clearly highlighting a successful extension of small volume indentation testing and analysis for mechanical properties assessment purposes that went beyond hardness and modulus alone just one year after the 1992 paper was published. Moreover, Oliver et al. also noted that additional improvements were also presented and went beyond the improved techniques prescribed by O&P just one year prior.

As time and attention progressed and evolved within the indentation-based research and development community, critical takes concerned with the OP data analysis and testing technique emerged as early as 1996 (if not earlier). Stated otherwise, a subset of the mechanical-properties-minded materials science and engineering world began to present alternative load-depth analysis procedures that were free of assumptions surrounding elastic material compatibility with depth-sensing indentation and even the OP approach in general. Other alternatives also noted concerns surrounding the reliance of OP upon a mean contact pressure definition of hardness, rather than that of energy-related principles (like that of the work of indentation), and even alongside a Meyers hardness perspective.

One of the early papers presenting such an alternative IIT load-displacement data analysis approach was rooted in the mechanical work of indentation, which may be thought of as a physical rendering of force and displacement at its' core and published in the mid-1990s by Gubicza et al. [39]. Beyond consideration of theory alone, Gubicza et al. also indirectly suggested that O&P's approach still performed just as well in achieving mechanical property evaluation compared with Gubicza et al.'s novel and semi-empirical depth-sensing indentation data analysis approach. Furthermore, Gubicza et al. stated that the hardness deduced agreed well with the OP method for many materials [39].

Still, Gubicza et al.'s critical take on the work of O&P was well-substantiated in so far as, ideally, elastic materials were of interest, for instance. However, in so far as the veracity of the work of Gubicza et al. is concerned, one ought to note that Gubicza et al.'s measurements and criticisms were levied using applied indentation pressures as high as 100 N, which resides within the macro-hardness regime rather than the micro-hardness and nano-hardness regimes initially affiliated with the OP technique as of 1992. Furthermore, the criticisms and critical takes levied by Gubicza et al., in so far as their findings were related and comparable to the findings of O&P, must be met with additional skepticism since they used a computer-controlled, Vickers indenter equipped, and hydraulic, mechanical testing device to perform depth-sensing indentation for hardness evaluation. Gubicza et al.'s elected use of such an IIT or indentation set-up and system controlled hydraulically rather than electromagnetically (in the case of the OP framework) prohibits genuine comparative analysis when considering the varied tip geometry, loading and depth recording sensitivities, and the like.

Beyond the initial considerations just detailed and discussed, a genuine pillar of the post-1992 OP-influenced era follows from the work of Field and Swain that was published in 1995. Indeed, Field and Swain's work has continued to garner traction through the present day because of Field and Swain's suspicion that sub-micrometer spherical and cyclic (i.e., partial loading-and-unloading cycles during a global loading-and-unloading process) nanoindentation testing could be used and extended into the realm of capturing mechanical flow curves and stress vs. strain plasticity phenomena. Such a window into mechanical flow behavior via Field and Swain's approach, or any other respectively similar approach concerned with indentation as the means of flow curve calculation, becomes particularly important when volumes of material in need of mechanical characterization cannot be characterized via traditional uniaxial tension or compression methods due to inherent size limitations. Accordingly, Field and Swain stated that spherical tip-based submicron indentation testing potentially housed the key to determining hardness, elastic modulus, representative stress vs. strain or mechanical flow curves, and strain hardening behaviors for size-limited material volumes [40].

As such, Field and Swain's representative spherical indentation stress vs. strain plot for a steel specimen, the physical condition of cono-spherical tips employed, and the degree of pile-up observed in one of the materials Field and Swain considered in their work. Advents since the time of Field and Swain rendered concerns surrounding the integrity of their approach since the size of the indents considered by Field and Swain suggests that Field and Swain surpassed the transitional limit of the spherical apex of their tips, which indicates that they unknowingly reported conical, rather than spherical, representative stress vs. strain curves at quite large indentation strains. In modern times, authors such as Sousa et al. in Refs. [41–43] and Leitner et al. in Refs. [44, 45] have consistently warned of the consequences surrounding the use of cono-spherical tips beyond their sphere-to-cone transition point, which results in a violation of Hertzian contact mechanics and geometrically defined stress-strain evolution.

Apart from the work of Field and Swain, the present section will also entertain further progress reportedly made pre-2002. In turn, attention is refocused upon another matter of depth-sensing indentation measurement that influences recorded load-displacement data and subsequently derived mechanical properties. Of particular focus at the respective point in time continued to consider the influences of pile-up and sink-in came into focus as findings suggested that improper accounting of pile-up can lead to the overestimation of hardness (since the area term associated with the denominator of Eq. (9) would be smaller than that corrected for pile-up, for example). Hence, Bolshakov and Pharr explored such matters in Ref. [46].

During the work by Bolshakov and Pharr, finite element analysis (FEA) of conical indentation of a variety of elastic-plastic materials were analyzed in-silico, enabling Bolshakov and Pharr to discover that underestimation of load-displacement curve derived contact areas could reach up to 60% when indentation-induced pile-up is large relative to indentation depth. Ultimately, Bolshakov and Pharr identified the ratio between h_f. and h_{max} measured parameters associated with recorded load-displacement data. One may take note of specimens wherein pile-up deserves more significant consideration than that of a correction factor, for example. Such a parameter was expressed as h_f/h_{max}, wherein a ratio less than or equal to 0.7 indicates that a material is not likely to be significantly affected by pile-up such that reasonable results are procurable via OP data analysis. Of course, the opposite was true when $h_f/h_{max} > 0.7$

6. Conclusion

The present chapter described the emergence of nanoindentation testing and analysis within the materials science and engineering literature. Emphasis was placed upon the origins of the dominant mode of submicron instrumented indentation testing and analysis (known as the Oliver-Pharr method). Specifically, detailed reconsideration and formulation of the Oliver-Pharr method was provided, followed by the industrial, engineering, manufacturing, and materials engineering R&D needs that were able to be addressed through the application of the Oliver-Pharr method. In detailing the emergence of Oliver and Pharr's approach to nanoindentation data analysis, the state of the art between the development of a nanoindenter in the early 1980s and formal publication of the Oliver-Pharr method in 1992 was presented. Next, several noteworthy and documented nanoindentation developments post-1992 were considered and contextualized. Continued consideration of the subsequent advancements, discoveries, innovations and attempts to realize nanoindentation's potential as an instrumented strength microprobe are described next in Part 2.

Author details

Bryer C. Sousa[1]*, Jennifer Hay[2] and Danielle L. Cote[1]

1 Department of Mechanical and Materials Engineering, Worcester Polytechnic Institute, Worcester, MA, USA

2 KLA Instruments (Oak Ridge, TN), KLA, Milpitas, CA, USA

*Address all correspondence to: bcsousa@wpi.edu

References

[1] Oliver WC, Pharr GM. An improved technique for determining hardness and elastic modulus using load and displacement sensing indentation experiments. Journal of Materials Research. 1992;7(6):1564-1583

[2] Doerner MF, Nix WD. A method for interpreting the data from depth-sensing indentation instruments. Journal of Materials Research. 1986; **1**(4):601-609

[3] Alisafaei F, Han CS. Indentation depth dependent mechanical behavior in polymers. Moshchalkov VV, editor. Adv Condens Matter Phys. 2015;**2015**: 391579

[4] Bulychev SI, Alekhin VP, Shorshorov MK, Ternovskij AP, Shnyrev GD. Determination of Young modulus by the hardness indentation diagram. Zavodskaya Laboratoria. 1975; **41**(9):1137-1140

[5] Poon B, Rittel D, Ravichandran G. An analysis of nanoindentation in linearly elastic solids. International Journal of Solids and Structures. 2008;**45**(24): 6018-6033

[6] Sneddon IN. The relation between load and penetration in the axisymmetric boussinesq problem for a punch of arbitrary profile. International Journal of Engineering Science. 1965; **3**(1):47-57

[7] Chandler H. Hardness Testing. 2nd ed. Materials Park, OH: ASM International; 1999. p. 210

[8] Fischer-Cripps AC. Multiple-frequency dynamic nanoindentation testing. Journal of Materials Research. 2004;**19**(10):2981-2988

[9] Tranchida D, Piccarolo S, Loos J, Alexeev A. Mechanical characterization of polymers on a nanometer scale through nanoindentation. A study on pile-up and viscoelasticity. Macromolecules. 2007;**40**(4):1259-1267

[10] Ebenstein DM, Pruitt LA. Nanoindentation of biological materials. Nano Today. 2006;**1**(3):26-33

[11] Hu C, Li Z. A review on the mechanical properties of cement-based materials measured by nanoindentation. Construction and Building Materials. 2015;**90**:80-90

[12] Oliver WC, Pharr GM. Measurement of hardness and elastic modulus by instrumented indentation: Advances in understanding and refinements to methodology. Journal of Materials Research. 2004;**19**(1):3

[13] Chen S, Liu L, Wang T. Investigation of the mechanical properties of thin films by nanoindentation, considering the effects of thickness and different coating–substrate combinations. Surface and Coating Technology. 2005;**191**(1):25-32

[14] Menčík J, Munz D, Quandt E, Weppelmann ER, Swain MV. Determination of elastic modulus of thin layers using nanoindentation. Journal of Materials Research. 1997;**12**(9): 2475-2484

[15] Hay J, Crawford B. Measuring substrate-independent modulus of thin films. Journal of Materials Research. 2011;**26**(6):727-738

[16] Li H, Chen J, Chen Q, Liu M. Determining the constitutive behavior of nonlinear visco-elastic-plastic PMMA thin films using nanoindentation and finite element simulation. Materials and Design. 2021;**197**:109239. ISSN: 0264-1275

[17] Saha R, Nix WD. Effects of the substrate on the determination of thin

film mechanical properties by nanoindentation. Acta Materialia. 2002; **50**(1):23-38

[18] Alaboodi AS, Hussain Z. Finite element modeling of nano-indentation technique to characterize thin film coatings. J King Saud Univ - Eng Sci. 2019;**31**(1):61-69

[19] Nix WD. Mechanical properties of thin films. Metallurgical Transactions A. 1989;**20**(11):2217

[20] Alexopoulos PS, O'Sullivan TC. Mechanical properties of thin films. Annual Review of Materials Science. 1990;**20**(1):391-420

[21] Oliver WC, Pharr GM. Nanoindentation in materials research: Past, present, and future. MRS Bulletin. 2010;**35**(11):897-907

[22] Pethicai JB, Hutchings R, Oliver WC. Hardness measurement at penetration depths as small as 20 nm. Philosophical Magazine A. 1983;**48**(4): 593-606

[23] Newey D, Wilkins MA, Pollock HM. An ultra-low-load penetration hardness tester. J Phys [E]. 1982;**15**(1):119-122

[24] Nishibori M, Kinosita K. Ultra-microhardness of vacuum-deposited films I: Ultra-microhardness tester. Thin Solid Films. 1978;**48**(3):325-331

[25] Tazaki M, Nishibori M, Kinosita K. Ultra-microhardness of vacuum-deposited films II: Results for silver, gold, copper, MgF2. LiF and ZnS. Thin Solid Films. 1978;**51**(1):13-21

[26] Nishibori M, Kinosita K. Ultra-microhardness of some vacuum-deposited films. Japanese Journal of Applied Physics. 1974;**13**(S1): A862B

[27] Fröhlich F, Grau P, Grellmann W. Performance and analysis of recording microhardness tests. Physica Status Solidi A: Applications and Materials Science. 1977;**42**(1):79-89

[28] Nix WD, Gao H. Indentation size effects in crystalline materials: A law for strain gradient plasticity. Journal of the Mechanics and Physics of Solids. 1998; **46**(3):411-425

[29] Oliver WC. Progress in the development of a mechanical properties microprobe*. MRS Bulletin. 1986;**11**(5): 15-21

[30] Doerner MF, Gardner DS, Nix WD. Plastic properties of thin films on substrates as measured by submicron indentation hardness and substrate curvature techniques. Journal of Materials Research. 1986;**1**(6):845-851

[31] Oliver WC, McHargue CJ, Zinkle SJ. Thin Film Characterization Using a Mechanical Properties Microprobe [Internet]. TN, USA: Oak Ridge National Lab.; 1987. Report No.: CONF-870388-5. Available from: https://www.osti.gov/biblio/6643878-thin-film-characterization-using-mechanical-properties-microprobe [Cited: 23 November 2021]

[32] Pethica JB, Oliver WC. Tip surface interactions in STM and AFM. Physica Scripta. 1987;**T19A**:61-66

[33] Oliver WC, Pethica JB. Method for continuous determination of the elastic stiffness of contact between two bodies [Internet]. US4848141A; 1989. Available from: https://patents.google.com/patent/US4848141A/en [Cited: 23 November 2021]

[34] Stone D, LaFontaine W, Alexopoulos P, Wu T, Li C. An investigation of hardness and adhesion of sputter-deposited aluminum on silicon by utilizing a continuous indentation test. Journal of Materials Research. 1988;**3**:141-147

[35] Loubet J, Georges J, Marchesini O, Meille G. Vickers indentation curves of

magnesium oxide (MgO). J Tribol-Trans Asme. 1984;**106**:43-48

[36] Pharr G, Oliver W. Nanoindentation of silver-relations between hardness and dislocation structure. Journal of Materials Research. 1989;**4**:94-101

[37] Pharr GM, Oliver WC. Measurement of thin film mechanical properties using nanoindentation. MRS Bulletin. 1992;**17**(7):28-33

[38] Oliver WC, Lucas BN, Pharr GM. Mechanical characterization using indentation experiments. In: Nastasi M, Parkin DM, Gleiter H, editors. Mechanical Properties and Deformation Behavior of Materials Having Ultra-Fine Microstructures. Dordrecht: Springer Netherlands; 1993. pp. 417-428

[39] Gubicza J, Juhász A, Lendvai J. A new method for hardness determination from depth sensing indentation tests. Journal of Materials Research. 1996; **11**(12):2964-2967

[40] Field JS, Swain MV. Determining the mechanical properties of small volumes of material from submicrometer spherical indentations. Journal of Materials Research. 1995; **10**(1):101-112

[41] Sousa BC, Sundberg KL, Gleason MA, Cote DL. Understanding the antipathogenic performance of nanostructured and conventional copper cold spray material consolidations and coated surfaces. Crystals. 2020;**10**(6): 504

[42] Sousa BC, Gleason MA, Haddad B, Champagne VK, Nardi AT, Cote DL. Nanomechanical characterization for cold spray: From feedstock to consolidated material properties. Metals. 2020;**10**(9). Available from: https://www.mdpi.com/2075-4701/10/9/1195

[43] Sundberg K, Sousa BC, Schreiber J, Walde CE, Eden TJ, Sisson RD, et al. Finite element modeling of single-particle impacts for the optimization of antimicrobial copper cold spray coatings. Journal of Thermal Spray Technology. 2020;**29**(8):1847-1862

[44] Leitner A. Advanced Nanoindentation Techniques for the Extraction of Material Flow Curves. University of Leoben. 2018

[45] Leitner A, Maier-Kiener V, Kiener D. Essential refinements of spherical nanoindentation protocols for the reliable determination of mechanical flow curves. Materials and Design. 2018; **146**:69-80

[46] Bolshakov A, Pharr GM. Influences of pileup on the measurement of mechanical properties by load and depth sensing indentation techniques. Journal of Materials Research. 1998;**13**(4): 1049-1058

Chapter 11

Toward an Instrumented Strength Microprobe – Origins of the Oliver-Pharr Method and Continued Advancements in Nanoindentation: Part 2

Bryer C. Sousa, Jennifer Hay and Danielle L. Cote

Abstract

Numerable advancements have afforded many benefits to nanoindenter system operators since the late 20th century, such as automation of measurements, enhanced load and displacement resolutions, and indentation with *in-situ* capabilities. Accordingly, the present chapter details how the Oliver-Pharr method of nanoindentation testing and analysis was adopted and relied upon as a framework that brought about widespread advancements in instrumented indentation testing. The present chapter introduces an emergent and theoretically consistent approach to assessing true stress–strain curves at a micromechanical scale using a flat-punch nanoindenter tip geometry and reliance upon Hollomon power-law plasticity and constitutive parameter fitting. Finally, a novel flat-punch nanoindentation testing method and approach to plasticity parameter analysis for metallic materials using nanoindentation systems can be implemented, bringing about an instrumented strength microprobe – a long sought-after tool.

Keywords: nanoindentation, instrumented indentation testing, elastic and plastic deformation, plasticity and strength, metallic materials, Hollomon plasticity, Oliver-Pharr method

1. Introduction

Similar to the work rendered and published by Bolshakov and Pharr in 1998, as discussed in Part 1, both Bolshakov and Pharr collaborated with Hay and Oliver in [1] to reformulate the Bolshakov-Pharr pile-up prediction ratio relation in terms of load–displacement curve slope-to-elastic contact stiffness ratio. Ultimately, Hay et al. found that S_l/S maintains a one-to-one ratio with h_f/h_{max} and E/σ_y, too, while also maintaining the capacity for direct measurement during testing, regardless of h_f or h_{max} inspection capabilities at a given facility. Indeed, **Figure 1** captures a few plots presented by Hay et al. when they formulated a pile-up constraint factor as a function of S_l/S as part of their 1998 research effort detailed in [1].

Nevertheless, post-OP article publication developments and research indeed focused upon relations beyond predicting and correcting for pile-up and relating

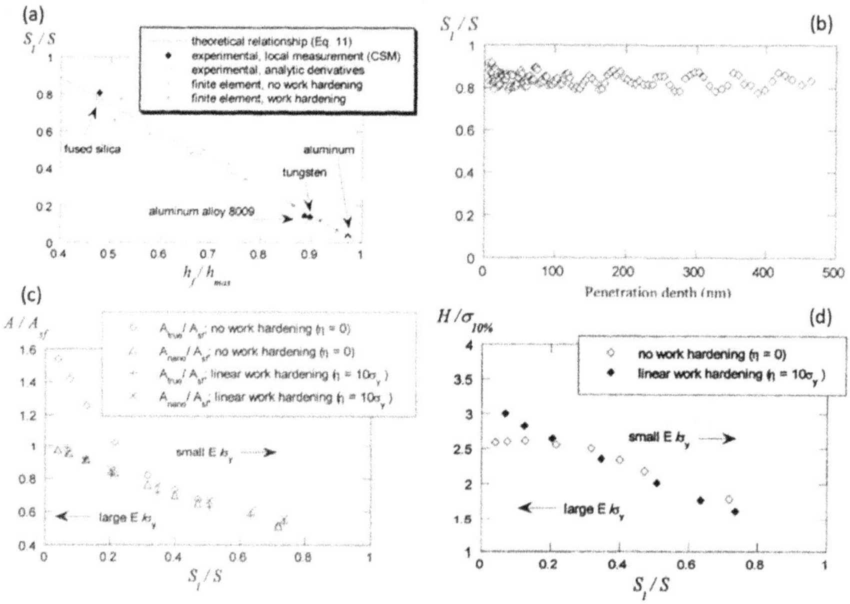

Figure 1.
(a) relation between S_l/S and h_f/h_{max}. (b) S_l/S vs. depth data obtained from testing fused silica [1]. (c) used the normalized contact area vs. S_l/S data to highlight that S_l/S ratios obtained at finite depths resulted in Anano's underestimation of actual contact area, which caused an overestimation of hardness as well as modulus. (d) dependency of the constraint factor as a function of S_l/S, wherein the constraint factor was found to be virtually independent of work-hardening when evaluated at large depths. Reproduced from [1].

the tendency of a material to pile up during nanoindentation to specimen properties, such as modulus, yield strength, work-hardening rate, and the like. Consequently, consider additional sources of influence and other externality-driven considerations of importance when performing and recording load–displacement testing and analysis. The work of Feng and Ngan highlighted how creep and thermal drift were both likely to influence modulus values recorded via instrumented indentation testing (IIT) or depth-sensing indentation since the OP method assumes that materials experience purely elastic recovery – an assumption that diverges from the physical reality underpinning the influencing factors upon recording and analyzing unloading segments of relevant nanoindentation data. More to the point, Feng and Ngan carefully constructed a simple scheme of creep effect corrections while measuring the modulus of elasticity. Furthermore, Feng and Ngan concurrently derived a way to nullify the thermal drift effects when measuring the modulus of elasticity [2]. Feng and Ngan achieved said derivations through the consideration of Al, Cu (111), and Ni_3Al (111) in [2].

In summation, Feng and Ngan showed that creep-based influences upon the compliance associated with the contact of an indenter tip with a given specimen during depth-sensing indentation and contact at the load–displacement point of initial unloading could be mathematically resolved and experimentally addressed via careful consideration of the unloading rate and max load hold times too. With the aforementioned in mind, one ought to recall that Feng and Ngan were mainly focused on the modulus of elasticity value modification or evolution when thermal drift and creep were not adequately understood and accounted for during nanomechanical, micromechanical, or both forms of indentation testing. However, unlike accounting for creep, which maintained significant improvements in measurement accuracy, a simple nullification protocol presented by Feng and Ngan ensured that thermal drift effects could be readily overcome and accounted for.

Around the same period, influences and relationships between those above-discussed phenomena and the ascertained mechanical properties gained traction through other researchers' perspectives. Indeed, in 1998, relations between the work of indentation, modulus of elasticity, and indentation hardness were considered by Cheng et al. [3]. Intriguingly, Cheng et al. resolved a proximal relation between the hardness-to-modulus proportionality and the proportionality between irreversible work of indentation and total work of indentation. As a result, Cheng et al. provided the nanomechanical research community with an alternative approach to estimating hardness and modulus via ITT systems equipped with conical or pyramidal indenter probes, as illustrated in part by **Figure 2** [3].

Still, fundamental explorations and evaluations of said relations need not stop with the work of Cheng et al.; rather, Hay et al. and others also maintained an early critical eye when inspecting underlying relations between properties and measured nanoindentation responses in [4–7].

Just as small-scale indentation-based and/or nanoindentation mechanical stress–strain curve evaluation captured the imagination of a subset of researchers cited in the previous subsection of the present manuscript, the period between 2003 and 2012 was also concerned with stress–strain behavioral insights via the use of nanoindentation systems. By 2003, Rodriguez and Gutierrez coupled an investigation of the ISE with a correlative analysis between tensile properties and nanoindentation-based behavior [8]. Rodriguez and Gutierrez invoked the use of the OP method of data analysis for performing sub-micron pyramidal

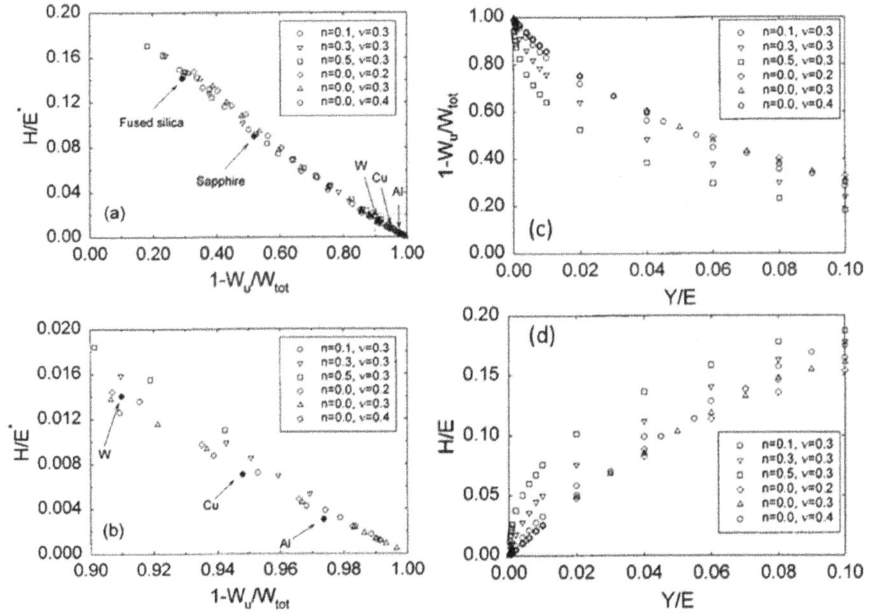

Figure 2.
Relationship between the proportional mechanical properties, including specimen hardness, measurable elastic modulus values, and various forms of indentation work, i.e., total, irreversible, etc. (a,b) Present the relationship between hardness/reduced modulus vs. the ratio of irreversible work-to-total work of indentation loading and unloading using computational finite element analysis and experimental data for various materials. (c) Presents the ratio of irreversible work-to-total work of indentation loading and unloading vs. the initial yield stress/modulus of elasticity, while (d) presents the hardness/reduced modulus vs. the initial yield stress/modulus of elasticity. Reproduced from [3].

nanoindentation while also invoking the strain gradient plasticity framework proposed by Nix and Gao for understanding the ISE affiliated with materials of various mesoscale mechanical tendencies such that the material flow stress could be compared with the bulk tensile properties of the very same specimens.

Therefore, when mathematically speaking, it stands to reason that plastic shear strain gradient-induced dislocations, or geometrically necessary dislocations, ρ^G, yields Eq. (1).

$$\rho^G \approx \delta\gamma \frac{1}{b\lambda} \tag{1}$$

wherein δ is a constant, γ is the plastic shear strain, b is the Burgers vector, and λ is the localized deformation field length scale relative to indentation size. At the same time, ρ^G and ρ^S, or the statistically stored dislocations, are cumulatively related to one another concerning the total dislocation density ρ^T via Eq. (2), such that.

$$\rho^T = \rho^G + \rho^S \tag{2}$$

Rodriguez and Gutierrez also invoked Taylor's relation, which expresses the mechanical flow stress σ of a material with the total dislocation density, such that Eq. (3) yields.

$$\sigma = \alpha M\mu b\sqrt{\rho^T} = \alpha M\mu b\sqrt{\rho^G + \rho^S} \tag{3}$$

whereinα is a constant, M is the Taylor factor, and μ is the elastic shear modulus.

Regrettably, Rodriguez and Gutierrez simply accepted the observation that the nanoindentation modulus of elasticity associated with the specimens studied, especially the steel-based subset of specimens, deviated from the macroscopic material's modulus. Rodriguez and Gutierrez did so by viewing the matter of deviating elastic modulus values through the lens of indentation-induced pile-up or sink-in influences. Ideally, a study claiming to be exploring a linkage between tensile stress–strain behavior and ISE-informed indentation data would have anticipated the effects introduced by pile-up and/or sink-in, which could have been experimentally addressed in part via confocal microscopy-corrected contact area values (given confocal microscopy's commercialization more than 20 years before the work of Rodriguez and Gutierrez). Rodriguez and Gutierrez also situated the then-debated direct relation between pile-up and material work-hardening.

Still, Rodriguez and Gutierrez correctly noted that grain-scale texture, surface roughness, and pile-up all influence nanoindentation data analysis, even going so far as to note that O&P, in 1992, did not correct for pile-up or sink-in effects. Of course, the lack of a method to account for pile-up or sink-in may limit O&P's approach as of 1992. However, one ought to note that O&P went on to incorporate such a correction factor during post-1992 refinements to the originally formulated and presented OP method.

Having just remarked that O&P made refinements to their original approach to analyzing nanoindentation derived data, one may consider one of the most noteworthy articles concerned with such refinements, which O&P has unsurprisingly presented since 1992. Said article by O&P is cited herein as [9]. As such, in their 2004 article, which was also published in the *Journal of Materials Research* (that is, the same journal that initially published their 1992 manuscript), O&P noted that improved surface contact identification was achievable via dynamic nanoindentation in comparison with static nanoindentation, which was first described in their appendix to the original 1992 article. O&P also noted the promising capacity of dynamic or CSM-based nanomechanical and nanoindentation-based micromechanical characterization to unveil mechanical properties as a function of depth throughout the entire loading and unloading process. Thirdly, O&P detailed how improvements could be integrated into nanomechanical testing systems to ensure more accurate area function calibration and load frame compliance assessment [9]. Each of the three improvements and refinements may be considered noteworthy when presented one by one, let alone all together in one document.

In any case, O&P continued their 2004 review of the progress made since their 1992 article by recognizing the fact that their method of nanoindentation testing and data analysis could be generalized to a greater degree and across more indenter geometries than previously thought. Importantly, O&P note that careful consideration must also be given to nanoindentation testing analysis of materials with reversible plasticity upon unloading since they assumed that only elastic displacements would be recoverable; yet, for those cases wherein materials partially unload plastically, O&P noted that FEM analysis had shown the effect to be virtually negligible for most monolithic materials. Of course, this remedied the previously noted concerns echoed by others in the field. Furthermore, in their original pedagogical formulation, O&P assumed that their expression for stiffness was sufficient; however, one significant improvement made by O&P by 2004 was the inclusion of a multiplicative β term as a function of the physical processes that may affect the value of the term such that Eq. (4) yields.

$$S = \frac{dP}{dh} = \frac{\beta 2E_r}{\sqrt{\pi}}\sqrt{A} \tag{4}$$

In so far as non-Berkovich, spherical nanoindentation load vs. displacement relations were concerned, O&P also demonstrated that at shallow indentation depths relative to the radius of a spherical tip, the load vs. depth relation may be expressed, in the elastic deformation regime, as Eq. (5), wherein.

$$P = {}^4\!/_3\sqrt{R}E_r\left(h - h_f\right)^{1.5} \tag{5}$$

such that differentiation of P relative to h, in conjunction with Eq. (5), yields Eq. (6), wherein.

$$S = \frac{dP}{dh} = 2\sqrt{R}E_r\left(h - h_f\right)^{0.5} \tag{6}$$

and Eq. (7) follows such that.

$$h_c = \frac{h_{max} + h_f}{2} \tag{7}$$

When proper substitution and algebraic manipulation are rendered, R is expressed as Eq. (8),

$$R = \left(\frac{1}{R_1} + \frac{1}{R_2}\right)^{-1} \tag{8}$$

while R_1 is the radius of the spherical indenter probe, and R_2 is the spherical hardness impression remaining after loading.

Interestingly, O&P also noted that the hardness obtained via spherical indenter tips was not necessarily equivalent to the hardness obtained using a Berkovich tip alongside their spherical nanoindentation contact mechanics analysis [9]. Work by Sousa et al. (see [10]) concurs with said hardness variability when a cono-spherical diamond indenter tip is dynamically applied to a material that is also tested via dynamic Berkovich nanoindentation according to OP methods in [10].

During the following year, Bei et al. examined the influence of using the contact area function, as determined via the OP calibration procedure, upon the nanomechanical phenomena known as pop-in events. The theoretical strengths are quantifiable from the analysis of said pop-in events [11]. Unlike sink-in or pile-up, which are related to either more or less material in contact with a given indenter tip during load vs. displacement cycling as part of indentation testing, pop-in events can be identified by abrupt bursts in depth at a given indentation load(s), which yield disjointed behavior within otherwise continuous load-depth data according to Bei et al.'s analysis. When Bei et al. presented their research, analysis, and findings, pop-in events were believed to signify the point of purely elastic to elastoplastic deformation during nanoindentation testing for both crystalline and amorphous material systems. Furthermore, pop-in events were believed to be connected with dislocation nucleation, although such associations have since been widely debated. Nevertheless, Bei et al. went on to detail the approach taken during their analysis, wherein finite element analysis was coupled with experimental load vs. displacement analysis to compare the effect of assuming a rounded spherical Berkovich tips' apex and the apex geometry gleaned from calibration procedures that had been documented in the past by O&P.

More specifically, the Virtual Indenter FEA simulation package from MTS Corporation (Knoxville, TN, USA), at the time, was utilized alongside MTS's nanoindentation testing system such that a diamond Berkovich indenter tip could be utilized in conjunction with a Cr_3Si single-crystal material system for their respective study. Interestingly, Bei et al. observed that the magnitude of the load at which a pop-in event occurred varied across crystallographic orientations and, after that, was rationalized through the lens of different resolved shear stress states and slip systems present in the material. Moreover, the findings by Bei et al. were consistent with their hypothesis that assuming a spherical apex, rather than a conical-spherical apex of a sharp pyramidal tip at sub 100 nm depths or so, could not provide an adequate geometrical description of the tip.

The influence of elastic anisotropy was not considered by Bei et al. as a potential influence upon indentation stress fields nor was the fact that sharp edges on Berkovich indenter tips were observable in residual deformation impressions considered. Such lack of consideration suggests that future work must address or overcome such problems, given that edge effects would undoubtedly change the stress field or stress state compared with a purely spherical geometry. Furthermore, because of the work of Bei et al. in 2005, the nanoindentation and nanomechanical characterization communities were made aware that simply assuming a spherical apex resulted in a maximum resolved shear stress that was overestimated by more than 41%. Lastly, Bei et al.'s experimental data correlated with the onset of the first pop-in event and dislocation nucleation once a material's theoretical shear strength was surpassed.

Pedagogical consideration of nanoindentation testing and load vs. displacement data analysis methods, especially those defined by O&P and widely accepted by the general nanomechanical and material characterization communities, continued into the mid-2000s. Another staple of the early sub-micron mechanical property investigation community, A. C. Fischer-Cripps was well aware of the general issues that users encountered through 2006 while attempting to perform nanoindentation testing, thus leading Fischer-Cripps to document common sources of error associated with performing nanoindentation testing such that ITT users who wished to utilize nanoindentation techniques could more readily and confidently do so in Ref. to [12].

Around the same time that Fischer-Cripps, O&P, and Bei et al. published their mid-2000s research, Troyon and Huang laid the additional groundwork for a correction factor, which was expressed by the β term presented in Eq. (4) by O&P in 2004; concurrent effects were appropriately accounted for during nanoindentation data analysis [13]. In doing so, Troyon and Huang detailed how radially inward and elastic displacements as well as indenter shapes deviations from a perfect cone geometry can concurrently be linked with one another in place of β, in the case of O&P in [6], or γ, in the case of work performed by Hay et al. in [4], alone. Soon after that, Troyon and Huang continued their research into the matter of a comprehensive multiplicative correction factor for mathematically detailing relations between contact area, modulus of elasticity, and unloading contact stiffness in another article too (see [14]). Moreover, [14] suggests that the Troyon and Huang methods should undergo continued analysis and extended applications; however, there exists no generalized advantage of their method with that of the OP method.

Like decades prior, the advancements made between 2013 and the present point centered upon pedagogical matters, the continued applicability of nanoindentation-based analysis, the development of a deeper understanding of plasticity and the mechanisms of materials, and more. Accordingly, we may first consider the work of Siu and Ngan, which was published in 2013 and suggested that the dynamic/CSM-based measurement method artificially induced sample strength modifications during oscillatory nanoindentation testing [15]. If true, such findings would generally question the integrity of CSM-based nanoindentation testing and the CSM-dependent advancements made within materials characterization as a field of study. Siu and Ngan made their case by coupling nanoindentation with microscopy (EBSD and TEM), *ex-situ*, while a Berkovich indenter tip geometry was utilized, CSM frequencies were varied, and a ductile, commercially pure, Al test specimen was used. Tentatively ignoring the veracity of the approach taken by Siu and Ngan for the time being and the correctness of their philosophical claims, the authors showed that CSM-induced errors could be decoupled from sample strain rate sensitivities. Interestingly, the work by Siu and Ngan resulted in the procurement of methodological findings and material deformation mechanism-based findings.

Stated otherwise, Siu and Ngan went beyond the realm of demonstration in 2013 and into the arena of physical mechanisms surrounding material plasticity. Siu and Ngan suggested that their unexpected observation of material softening could be interpreted through the lens of variable pressure during oscillatory cycling, enabling stress relief due to elastic recovery during the unloading half cycle, wherein the material was in an elastoplastic state. Moreover, Siu and Ngan argued that those mentioned above, in turn, induced dislocation motion reversals, which resulted in dipole dislocation annihilation, decreased dislocation density, and dislocation motion-induced subgrain formation. Finally, although Siu and Ngan noted that the OP method of hardness determination appeared to be flawed when CSM capability was enabled, Siu and Ngan also conceded that the influence of nanometric oscillations upon nanometer length scale deformation of metals was unknown at the time [15].

Nevertheless, Al was not the only material or work concerned with related or similar matters of relevance following their Al-based 2013 article. In fact, during the same year, Siu and Ngan extended such oscillation-induced strength modification effects to Cu and Mo, too [16]. Furthermore, motivated by their original findings detailed in [15], Siu and Ngan stated that oscillation-induced softening of Al was intrinsically intertwined with enhanced annihilation of dislocations and formation of sub-grains due to the simultaneous imposition of oscillatory stresses [16].

Accordingly, such an interpretation and distillation of their earlier findings were consistent with their respective motivation underlying the use of Cu and Mo to go beyond proof-of-concept demonstration and toward the realm of generalized phenomena for a particular class of metallic materials.

While Siu and Ngan continued to build upon their 2013 research elsewhere [17, 18], other researchers garnered a maintained interest in the topic [19–21]. Additional oscillatory-induced and CSM measurement methods incurred errors studied by varying nanoindentation research and development community members. Inspired by the 2006 publication by Durst et al. [22], as well as the work of Cordill et al. in 2009 [23], alongside Pharr et al.'s more recent account of Durst et al.'s findings in [24] via modulus-to-hardness material ratios in addition to unloading curve assumptions and their effect on evaluated stiffness values, and Vachhani et al.'s reported outcome departures during dynamic nanoindentation testing as a function of harmonic frequency in [25], Merle et al. set out to advance a generalized understanding surrounding Vachhani et al.'s findings in [26].

Consequently, Merle et al. found that caution must be exercised appropriately when evaluating materials with high modulus-to-hardness ratios because selected harmonic parameters cause notable contact stiffness underestimation, which directly influences the resultant modulus of elasticity recorded when high loading rates are employed [26]. In doing so, Merle et al. were able to identify the culprit responsible for such oscillation-induced behavior in terms of a biased phenomenon associated with lock-in amplifier signal processing. Furthermore, Merle et al. illustrated how phase angle signal data could indicate the occurrence of said oscillatory artifacts during dynamic nanoindentation testing.

Turning our attention to advancements made in so far as the utility of the work of indentation is concerned, per modernized understanding of nanoindentation as a field of study, the 2013 publication by Jha et al. is considered next [27]. Jha et al. executed extensive nanoindentation modeling via finite element analysis methods to probe elasticity and elastoplastic behavior to garner insights into the total work of indentation and the elastic work of indentation from load–displacement data. Jha et al. found that the aforementioned work parameters could characterize the mechanical response of materials under indentation [27]. Based on the work of Jha et al., one may consider the effort as a means of better appreciating the concept of the work of indentation (which was already discussed in the previous subsections) and the agreement shared between the contact depths obtained for Berkovich and spherical indenter tips coupled with the OP method and that obtained via elastic energy constant-based contact depth determination methods in [27].

In much the same way the nanoindentation research and development concerned with the notion and concept of the work of indentation extended into the current decades' advancements, nanoindentation stress–strain curve evaluation was undoubtedly a focal point of research activity to date too. Such sentiment holds in so far as the influence of CSM measurement parameters is considered to a greater extent at the same time too. However, by keeping with the notable research published in 2013 concerning said matters of inquiry, we may first consider the work of Bobzin et al. [28]. In 2013, Bobzin et al. attempted to address the standing need for prospective nanoindentation-based mechanical flow curve derivation methods via iterative comparative analysis of FEA computations. They experimentally obtained load vs. displacement curves coupled with adaptive plastic behavior model parameter identification. Interestingly, Bobzin et al. echoes less-utilized stress–strain nanoindentation analysis by Juliano et al. [29], such that reference stress and reference strain values are measurable from load-depth data and curves via Eq. (9) and Eq. (10), respectively. Thus, the following mathematical relation for reference stress was invoked by Bobzin et al., such that.

$$\sigma = \frac{0.9P}{\pi a^2}\left(\frac{h_t}{h_c} - 1\right) \tag{9}$$

while the following mathematical relation for reference strain was also provided by Bobzin et al., such that.

$$\varepsilon = \frac{\sigma}{1 - v^2}\left(\left(\frac{4}{2.7P\left(h_t h_c^{-1} - 1\right)}\left(2h_t - 2h_c\frac{1}{\left[2h_c\left(a^2 + h_c^2\right)^{-1}\right]^{\frac{1}{3}}}\right)\right)^{\frac{3}{2}} - \frac{1 - v_i^2}{E_i}\right) \tag{10}$$

and linked together via the Johnson-Cook model, as given in the following expression, depicted herein as Eq. (11), such that.

$$\sigma = A + B\varepsilon^n \tag{11}$$

Just a few years after the work of Jha et al., Bobzin et al., Siu and Ngan, and Merle et al., which are detailed above, nanoindentation stress vs. strain evaluation methods, other than Bobzin et al.'s FEA-supported protocol, were applied to the mechanical characterization and property evaluation tasks surrounding alloyed Ti64 materials in [30]. Though the spherical nanoindentation-derived effective stress vs. strain-curve methodologies and protocols were established before the work of Weaver and Kalidindi, which is currently under consideration, Weaver and Kalidindi clearly and concisely noted their intended objectives and motivation behind their 2016 manuscript in its' introductory section. Weaver and Kalidindi provided the materials development and design research and engineering communities with a referable case study that ideally captured the high-throughput nature and feasibility of such material characterization in so far as metal systems were concerned, including Ti64, which maintains microstructural complexity concerning site-specific, microstructural, and local length-scale dependencies of measured material response too [30].

Moreover, the protocols above of relevance, which include Kalidindi and Pathak's approach, as detailed in the previous section of the present literature review, were relied upon throughout most of the analysis and mechanical characterization performed in Weaver and Kalidindi's 2016 study. Building upon the research presented in 2016, as well as the advancements and protocol developments by Kalidindi and close colleagues across Drexel University of the Georgia Institute of Technology, as well as others, Weaver, and others from the same network relating Kalidindi, Pathak, and Weaver with one another, continued to extend the range of spherical nanoindentation stress vs. strain exploration and data analysis. Intriguingly, Weaver et al. invoked CSM spherical nanoindentation stress–strain analysis to probe and measure the evolution of mechanical properties of He, W, and (He+W) ion-irradiated tungsten at a granular level in [31].

Nevertheless, an unexpected observation of significance for those utilizing OP analysis modalities was the secondary finding by Weaver et al. that ion-irradiated hardness values obtained via OP and Berkovich indentation and Weaver et al.'s spherical dynamic nanoindentation approach can be directly compared with one another. Still, Weaver et al. are not the only modern researchers concerned with spherical indentation and indentation-based stress–strain characterization and data analysis. Instead, Xiao et al. in 2019 unveiled a mechanistic model for spherical nanoindentation stress vs. strain relations, which intertwined three principal deformation mechanisms of relevance [32]. More to the point, said principal deformation mechanisms invoked by Xiao et al. included the indentation size effect and, therefore, the matter of geometrically necessary dislocations, followed by irradiation hardening and were concluded in terms of strain-softening affected by the removal of defects. The performance of Xiao et al.'s model for non-irradiated reference materials is compared to experimentally measured values and the model's predictions of indentation stress versus indentation strain. Additionally, this performance is compared to an alternative model previously reported in the literature before Xiao et al.'s publication.

Beyond the work of Xiao et al. cited above, additional developments of relevance include that detailed in 2018 by Jin et al., wherein Jin et al. claimed to have formulated a model for the quantifiable linkage of incipient irradiation damage via nanoindentation pop-in phenomena analysis in [33]. After that, in 2019, the near-surface nanoindentation response of ion-irradiated FCC metals was studied by way of strain gradient plasticity mechanisms, which were modified to account for irradiation effects [34]. At the same time, Xiao et al. extended their 2019 single-crystal irradiation studies into the realm of polycrystalline steel specimens [35].

On the other hand, just a couple of years prior to the work of Xiao et al. in 2019, Kumar et al. attempted to develop a nanoindentation-based means of evaluating argon-ion irradiation-induced hardening of a ferritic and martensitic dual-phase steel in [36]. As a result, Kumar et al. found changes in nanoindentation hardness of dual-phase steel because irradiation follows a power-law relation dependent upon irradiation dosage [36].

Returning to the realm of spherical nanoindentation stress vs. strain curve analysis through Weaver, Pathak, and Kalidindi, one may note [37], wherein Khosravani et al. focused on spherical nanoindentation as a means of characterizing two hierarchical and martensitic FeNiC steel systems. Interestingly, Khosravani et al. found that pop-in phenomena could be linked to dislocation and lath boundary interaction, followed by dislocation transmission through the boundary during nanoindentation loading. Beyond the critical examination that Khosravani et al. provided regarding how the measured properties may be related to uniaxial tensile test counterparts, Khosravani et al. also quantified the mechanical behavior of lath martensite phases across length scales and various indenter tip radii. Khosravani et al. reportedly observed minor indentation size effects across variable tip radii and nanometric length-scale plasticity and strength domination by nanostructured defects.

Unsurprisingly, steel continues to capture the attention of nanoindentation researchers and material science and engineering community members to date. For example, in 2020, Massar et al. applied dynamic Berkovich nanoindentation hardness and modulus of elasticity measurements, according to the OP method, for recycled battlefield scrap steel powders and cold spray-processed material consolidations in [38]. In addition to Massar et al., continued consideration of the use of nanoindentation to characterize steels in the 2010s unveils the work of Pham and Kim, which was published in 2015, wherein the authors identified the modulus of elasticity and nanoindentation hardness values associated with an SM490 steel weld zone via statistical data analysis [39]. Furthermore, Yang et al. also utilized statistical data analysis to quantify the transformation kinetics of bainite phase formation within an austempered steel [40]. Still, nanoindentation for such steel specimen characterization is also discussed elsewhere, including the following references of note: [41–45], to name a few.

One ought to also consider the recent work of Ruiz-Moreno et al., published in 2020, as another notable nanoindentation-related research article [46]. Ruiz-Moreno et al. performed nanomechanical characterization over a range of temperatures. Ruiz-Moreno et al. demonstrated how indentation hardness of a P91 system could be measured under ambient conditions and at elevated temperatures (873.15°

K) for transient property inspection via quasi-static and dynamic nanoindentation testing, as shown in **Figure 3**. In addition to the range of transient properties explored at elevated temperatures and ambient temperature, Ruiz-Moreno et al. ventured into pedagogical consideration surrounding assessing spherical

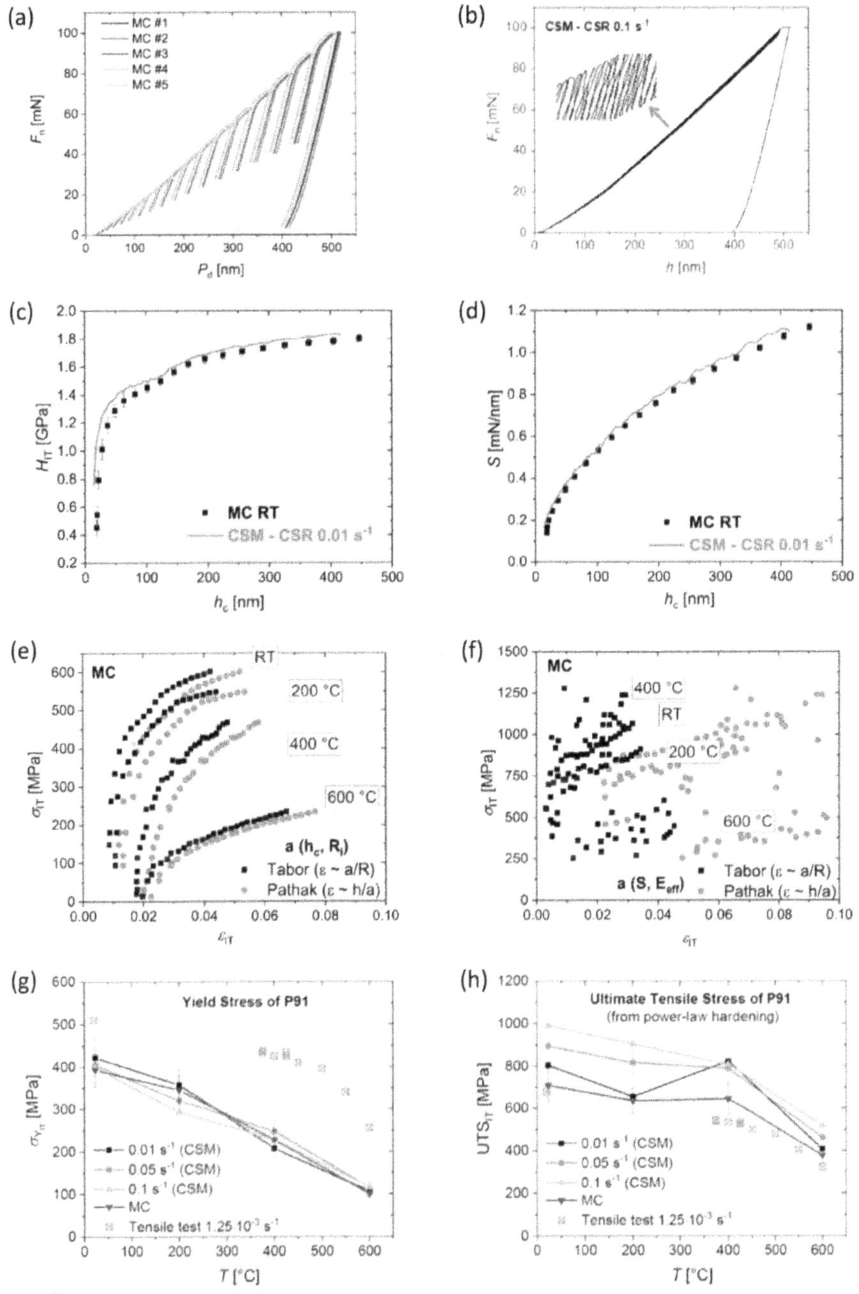

Figure 3.
Cyclic or quasi-static nanoindentation load-depth curves are presented in (a), while a dynamic nanoindentation load vs. depth curve is presented via (b). Indentation stress–strain curves as a function of testing temperature are presented in (e,f). Additional details shown in (c,d,g,h) can be understood through consulting [46], which is the reference the present figure is adopted and reproduced from.

nanoindentation stress vs. strain curves. Conditions, Ruiz-Moreno et al. noted that consequences existed when adhering to the school of thought surrounding and underpinning the work of Weaver, Pathak, Kalidindi, and the like [46].

2. Actualization of a frustrum instrumented strength microprobe

As was discussed above, many have endeavored to use a spherical nanoindenter tip probe to measure small-scale stress–strain curves with limited or variable degrees of success [47]. However, spherical nanoindentation includes several practical difficulties because the uncertainty in the contact area is generally more significant for spheres than for other indenter shapes (see the variation in an as-manufactured cono-spherical tip geometry obtained from a leading tip supplier for more nuanced appreciation, as shown in **Figures 4–7**, which are adopted from [10].

Furthermore, as the contact area grows during spherical nanoindentation testing, the volume of tested material also grows, gradually and continuously incorporating virgin material into the test. Thus, both the material and strain change concurrently. However, with a flat punch, the contact area is well known and fixed as the area of the punch face. Because the contact area is fixed, the volume of the tested material is roughly constant throughout the entirety of the test. Thus, the present section demonstrates how such fixed flat-punch indentation testing conditions or parameters enable microscopic true stress–strain relationships of a wide range of metallurgical materials to be probed using Hay et al.'s emergent protocol (a protocol that has ultimately brought the nanoindentation community one step closer to realizing the long sought-after Instrumented Strength Microprobe).

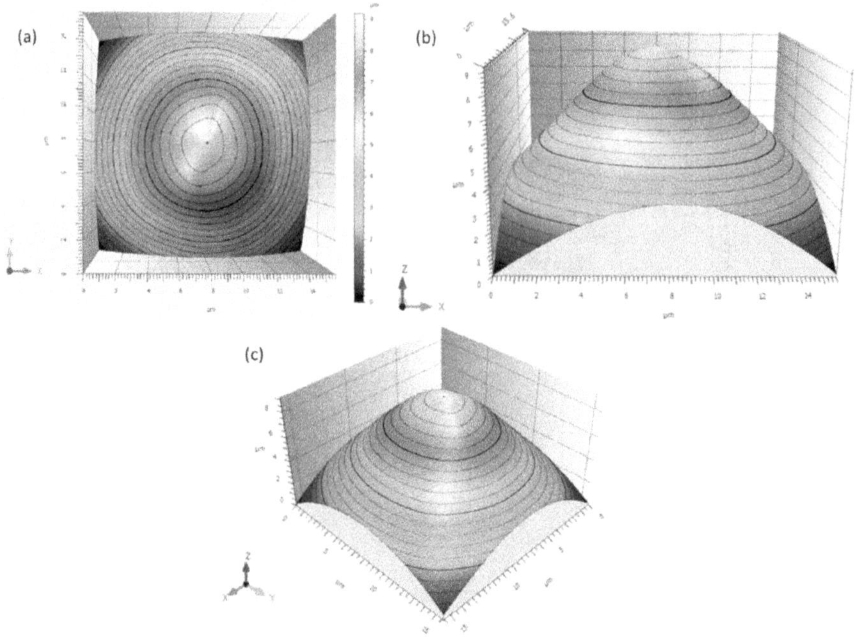

Figure 4.
Renderings of the imperfect tip geometry obtained when a cono-spherical tip was purchased from Micro-star technologies: (a–c) have been normalized in the x, y, and z directions. (a) Degree of deviation of the actual tip geometry from an ideally spherical tip at the apex of the indenter probe. Notice the scale bar in microns to the right of (a), which topographically signifies the distance from the point of Cartesian coordinate origin parallel to an x-y plane that is orthogonal to the z-axis. Reproduced from [10].

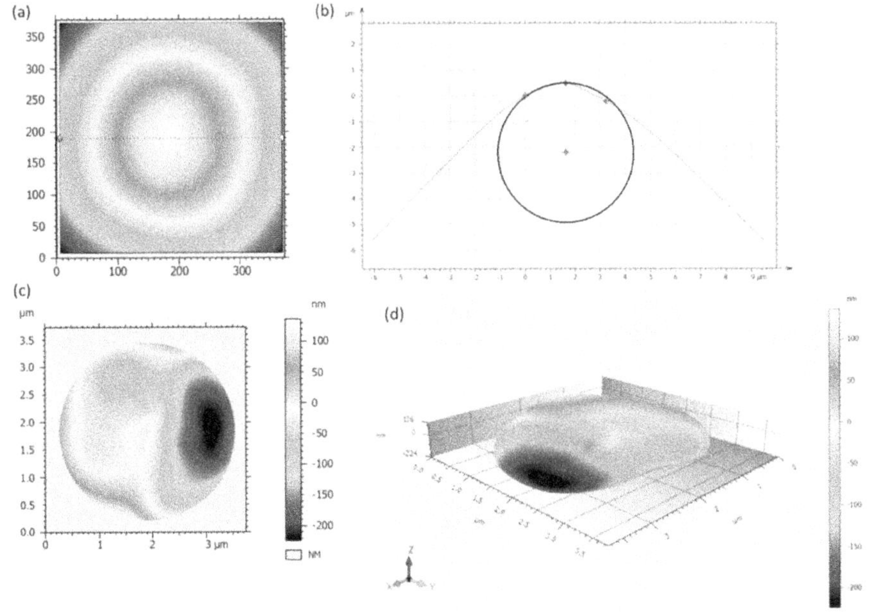

Figure 5.
(a) location of the extracted tip profile, obtained through the very tip of the probe's apex. (b) a circle with a nominal radius of 2.71 μm fit to the presented profile. The circle in (b) is positioned such that it is coincident with the tip of the apex and parallel to the z-axis. Notice the deviation from the imposed circle in (b) by the profile between approximately 1.5 and 3.5 μm on the horizontal axis. (c-d) deviation of the true surface from an ideal sphere of a 2.71 μm radius. (c-d) is normalized in the x- and y-directions with vertical distortion in the z-direction to capture the deviation via colored contour plotting. Reproduced from [10].

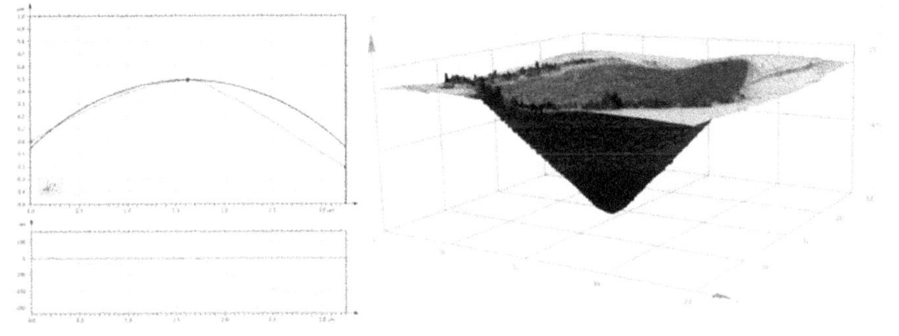

Figure 6.
Residual indentation imprint on a single crystal of commercially pure Al was obtained using 3D confocal microscopy-based analysis. The upper graphical rendering illustrates the true surface profile at the spherical region in the blue line versus the ideal surface profile for a spherical tip with a 2.71 μm radius in the black line. The plot at the bottom of this figure captures the surface profile deviation from a 2.71 μm nominal radius circle. Reproduced from [10].

The microscopic stress–strain relationships were measured by nanoindentation using a new method [48]. This method required Young's modulus as input, and it returned the yield point, true stress–strain ordered pairs beyond the point of complete contact, and coefficients K and n of the best power-law fit to the post-yield behavior Eq. (12):

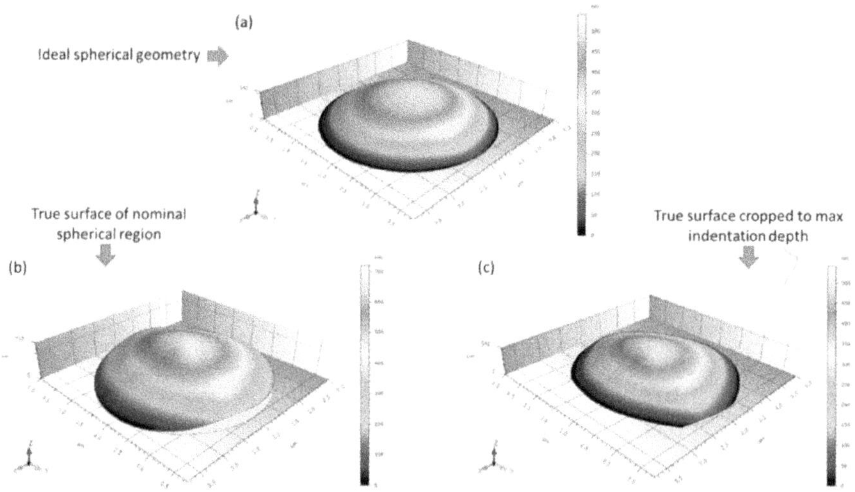

Figure 7.
Even though the maximum indentation depth associated with a nanoindenter tip of 2.71 μm is 794 nm, the max cut-off depth before the transition from spherical shape to a conical geometry was set at 542 nm to showcase the spherical geometric profiles readily. Therefore, at 542 nm away from the apex of the perfectly spherical tip, the ideal spherical geometry is presented in (a). (b) Unveils the true surface profile associated with the tip from Micro-star technologies up to 794 nm from the apex. For comparison with (a), (c) captures the actual surface cropped to the same vertical dimension as the ideal surface, highlighting the deviation. Renderings are normalized in the x, y, and z directions. Reproduced from [10].

$$\sigma = K\varepsilon^n \tag{12}$$

With a KLA iMicro nanoindenter, a 90-degree frustum, having a flat circular end with a diameter of around 10 μm, can be pressed into a given target material at a prescribed strain rate and to a depth of 2 μm while continuously measuring load, P; penetration depth, h; and elastic contact stiffness, S, with the latter being measured by a small, superimposed oscillation [49, 50]. The strain is calculated as Eq. (13) for each point acquired during loading beyond the point of complete contact:

$$\varepsilon = \left(\frac{2}{\pi}\right)\frac{h}{a} \tag{13}$$

wherein a is the contact radius, calculated as the radius of the indenter tip at a distance h from the frustum face along the indenter axis, or the following relation, shown in Eq. (14):

$$a = a + htan\psi \tag{14}$$

with ψ representing the half-included angle of the cone. For each point acquired during loading beyond the point of complete contact, the stress σ was calculated in proportion to the mean pressure of the contact p_m, as Eq. (15) below:

$$\sigma = \xi p_m \tag{15}$$

with the constant proportionality ξ calculated as a linear function of the parameter S^* according to Eq. (16) as follows,

$$\xi = 0.3969(S^*) + 0.3218 \tag{16}$$

where S^* is calculated as S_L/S or the local slope of the unloading curve dP/dh divided by the elastic contact stiffness at the same point. Finally, the expression for strain rate is given as Eq. (17):

$$\varepsilon' \equiv \left(\frac{2}{\pi}\right)\frac{h'}{a} \tag{17}$$

Logically, the target strain rate ε' is prescribed by a given user, and the system calculates and imposes the necessary penetration rate h' to achieve the target strain rate. In any case, the point of complete contact between the punch face and test surface was determined as the point where measured contact stiffness first exceeded the expected value based on the known reduced modulus for the material E_r [51, 52], which is given in Eq. (18):

$$S > 2E_r a \tag{18}$$

To determine the yield point, the stress–strain ordered pairs acquired beyond the point of complete contact fit the power-law form of Eq. (12); the yield point for each test was determined as the point of intersection between this power-law fit and the linear part of the stress–strain curve (as determined by the input Young's modulus). Thus, by the end of each test, the whole true stress–strain curve was available by automatically patching together these three segments into a whole: (i) from the origin to the yield point: the linear part (generated by the input Young's modulus); (ii) from the yield point to the point of complete contact: an extrapolation of the power-law fit to data from segment (iii); and (iii) from the point of complete contact to the end of loading: measured stress–strain as calculated by Eq. (13) and Eq. (14), respectively.

That said, the relationship for strain that is defined in Eq. (13) was derived by considering elastic contact between a frustum and a test surface, with the load and depth related through the reduced modulus E_r, as Eq. (19) [51].

$$P \equiv 2E_r a h \tag{19}$$

Dividing both sides of Eq. (19) by the contact area gives an expression for mean pressure, Eq. (20):

$$\frac{P}{\pi a^2} = p_m = (2/\pi)E_r(h/a) \tag{20}$$

If we define the strain as in Eq. (13), then Eq. (20) becomes Eq. (21), such that.

$$p_m = E_r \varepsilon_i \tag{21}$$

which is analogous in form to the stress–strain relation used to comprehend the elastic part of a uniaxial tension or compression test. Because one definition of strain should befit both elastic and plastic phases of the test, we used the definition of strain expressed by Eq. (13) for all testing, and by extension, the definition of strain rate expressed by Eq. (17).

The linear function for inferring the constant of proportionality ξ from S^* Eq. (16) was determined using extensive (90+) FEM simulations of flat-punch indentations into materials with systematically varied degrees of plasticity. The extent of plasticity for each finite-element simulation was captured by a single value of the parameter $S^* \equiv S_L/S$ [1, 53]. For each simulation, S_L was determined as the slope of the loading curve prior to peak penetration; the contact stiffness S was determined as the slope of the simulated load-depth relation at the onset of unloading. (Experimentally, S_L and S are available throughout loading, but S is only available at the onset of unloading.) For fully elastic simulations, S^* had a value of unity because the loading and unloading curves coincided. For the most plastic simulations, S^* approached zero as the unloading curve was nearly vertical. Thus, the domain for S^* is zero (fully plastic) to unity (fully elastic).

To determine the precise functional form of $\xi = f(S^*)$, each finite-element simulation was analyzed as follows: (i) the parameter's value S^* was calculated; (ii) strain was calculated according to Eq. (13); (iii) the true stress value σ was calculated from the input stress–strain relation as the stress at the strain calculated in step (ii); and (iv) the value of ξ was calculated by dividing the true stress (step iii) by the mean indentation pressure p_m. Finally, using the results of all simulations, it was found that ξ depended linearly on S^*, so long as $S^* < 0.8$; Eq. (16) is the best fit for this linear relation. We note here that the intercept of Eq. (16), which indicates the value of ξ for the case of full plasticity ($S^* \to 0$), is very close to the scaling factor of 1/3 determined by David Tabor [54] and further verified by many others through finite-element analysis [55] and experiments [56].

Figure 8 presents the true stress–strain curves obtained for multiple flat-punch indents via the method described herein and applied to commercially pure Ti. The average yield stress obtained was 270 ± 50 MPa, and individual test data from twelve measurements using this method agreed with 240 MPa as measured via tensile testing while also being well within the upper and lower bounds of yield strength for Ti reported within the relevant literature. While the scatter within the data was high, the microstructure of the pure Ti system resulted in relatively large grains, which would allow for a single test location to be constrained to single as well as randomly oriented grains; further information about Hay's original work and the KLA-patented flat-punch stress vs. strain analysis technique can be found in [57], while the present authors have detailed recent applications and extensions of the method in [58].

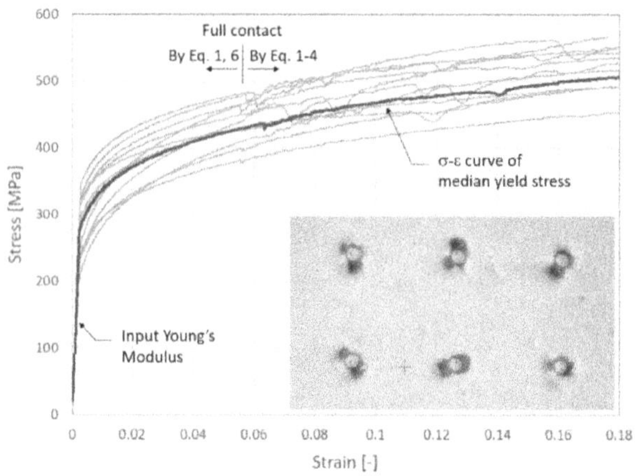

Figure 8.
True stress–strain for a Ti metal system characterized using the method formulated further herein and Hay's patented and emergent flat-punch nanoindentation technique. Reproduced from [57].

3. Conclusions

Building off the discussion provided in Part 1, the present chapter continues the consideration of advancements within the theory and practice of small-volume instrumented indentation testing and nanoindentation testing and analysis enabled via the advent of the Oliver-Pharr method. Advancements presented within the literature following Oliver and Pharr's 1992 research article (and discussed herein) focused upon contact area estimation, pile-up phenomena, stiffness, creep-correction, thermal drift, relations between measured properties and work of indentation terms, load–displacement curve, pop-in phenomena, the application of strain gradient plasticity, and more. Thereafter, consideration of prior work presenting spherical nanoindentation testing as a means of assessing indentation stress–strain curves of metallic materials (in particular) was discussed, followed by transient material property assessment with high-temperature nanoindenter systems and the effects of continuous stiffness measurement testing procedures and their influence upon recorded results. In conclusion, the present chapter is finalized through the introduction of an emergent and theoretically consistent approach to assessing true stress–strain curves at a micromechanical scale using a flat-punch nanoindenter tip geometry and reliance upon Hollomon power-law plasticity and constitutive parameter fitting, therefore detailing the long sought-after ability to utilize nanoindentation testing as an instrumented strength microprobe.

Author details

Bryer C. Sousa[1*], Jennifer Hay[2] and Danielle L. Cote[1]

1 Department of Mechanical and Materials Engineering, Worcester Polytechnic Institute, Worcester, MA, USA

2 KLA Instruments (Oak Ridge, TN), KLA, Milpitas, CA, USA

*Address all correspondence to: bcsousa@wpi.edu

References

[1] Hay JL, Oliver WC, Bolshakov A, Pharr GM. Using the ratio of loading slope and elastic stiffness to predict pile-up and constraint factor during indentation. MRS Proceedings. 1998; **522**:101

[2] Feng G, Ngan AHW. Effects of creep and thermal drift on Modulus measurement using depth-sensing indentation. Journal of Materials Research. 2002;**17**(3):660-668

[3] Cheng YT, Cheng CM. Relationships between hardness, elastic modulus, and the work of indentation. Applied Physics Letters. 1998;**73**(5): 614-616

[4] Hay JC, Bolshakov A, Pharr GM. A critical examination of the fundamental relations used in the analysis of nanoindentation data. Journal of Materials Research. 1999;**14**(6): 2296-2305

[5] Fischer-Cripps AC. A review of analysis methods for sub-micron indentation testing. Vacuum. 2000; **58**(4):569-585

[6] McElhaney KW, Vlassak JJ, Nix WD. Determination of indenter tip geometry and indentation contact area for depth-sensing indentation experiments. Journal of Materials Research. 1998; **13**(5):1300-1306

[7] Martin M, Troyon M. Fundamental relations used in nanoindentation: Critical examination based on experimental measurements. Journal of Materials Research. 2002;**17**(9): 2227-2234

[8] Rodriguez R, Gutierrez I. Correlation between nanoindentation and tensile properties: Influence of the indentation size effect. Materials Science and Engineering A. 2003; **361**(1):377-384

[9] Oliver WC, Pharr GM. Measurement of hardness and elastic modulus by instrumented indentation: Advances in understanding and refinements to methodology. Journal of Materials Research. 2004;**19**(1):3

[10] Sousa BC, Gleason MA, Haddad B, Champagne VK, Nardi AT, Cote DL. Nanomechanical characterization for cold spray: From feedstock to consolidated material properties. Metals. 2020;**10**(9). Available from: https://www.mdpi.com/2075-4701/10/9/1195

[11] Bei H, George EP, Hay JL, Pharr GM. Influence of indenter tip geometry on elastic deformation during Nanoindentation. Physical Review Letters. 2005;**95**(4):045501

[12] Fischer-Cripps AC. Critical review of analysis and interpretation of nanoindentation test data. Surface and Coating Technology. 2006;**200**(14): 4153-4165

[13] Troyon M, Huang L. Correction factor for contact area in nanoindentation measurements. Journal of Materials Research. 2005;**20**(3): 610-617

[14] Troyon M, Huang L. Comparison of different analysis methods in nanoindentation and influence on the correction factor for contact area. Surface and Coating Technology. 2006; **201**(3):1613-1619

[15] Siu KW, Ngan AHW. The continuous stiffness measurement technique in nanoindentation intrinsically modifies the strength of the sample. Philosophical Magazine. 2013; **93**(5):449-467

[16] Siu KW, Ngan AHW. Oscillation-induced softening in copper and

molybdenum from nano- to micro-length scales. Materials Science and Engineering A. 2013;**572**:56-64

[17] Cheng B, Leung HS, Ngan AHW. Strength of metals under vibrations – Dislocation-density-function dynamics simulations. Philosophical Magazine. 2015;**95**(16–18):1845-1865

[18] Siu K, Liu H, Ngan A. A universal law for metallurgical effects on acoustoplasticity. Materialia. 2019;**5**: 100214

[19] Golovin YI, Korenkov V, Razlivalova S. Effect of low-amplitude load oscillations on stiffness and hardness for Al and W in loaded Nanocontacts. Bulletin of the Russian Academy of Sciences: Physics. 2018; **82**(9):1180-1186

[20] Kanders U, Kanders K. Nanoindentation response analysis of thin film substrates-II: Strain hardening-softening oscillations in subsurface layer. Latvian Journal of Physics and Technical Sciences. 2017;**54**(2):34-45

[21] Kanders U, Kanders K. Strain hardening-softening oscillations induced by nanoindentation in bulk solids. ArXiv Prepr ArXiv160909791. 2016

[22] Durst K, Backes B, Franke O, Göken M. Indentation size effect in metallic materials: Modeling strength from pop-in to macroscopic hardness using geometrically necessary dislocations. Acta Materialia. 2006; **54**(9):2547-2555

[23] Cordill M, Lund M, Parker J, Leighton C, Nair A, Farkas D, et al. The Nano-jackhammer effect in probing near-surface mechanical properties. International Journal of Plasticity. 2009; **25**(11):2045-2058

[24] Pharr GM, Strader JH, Oliver W. Critical issues in making small-depth mechanical property measurements by nanoindentation with continuous stiffness measurement. Journal of Materials Research. 2009;**24**(3):653-666

[25] Vachhani S, Doherty R, Kalidindi S. Effect of the continuous stiffness measurement on the mechanical properties extracted using spherical nanoindentation. Acta Materialia. 2013; **61**(10):3744-3751

[26] Merle B, Maier-Kiener V, Pharr GM. Influence of modulus-to-hardness ratio and harmonic parameters on continuous stiffness measurement during nanoindentation. Acta Materialia. 2017; **134**:167-176

[27] Jha KK, Zhang S, Suksawang N, Wang TL, Agarwal A. Work-of-indentation as a means to characterize indenter geometry and load–displacement response of a material. Journal of Physics Applied. 2013;**46**(41): 415501

[28] Bobzin K, Bagcivan N, Theiß S, Brugnara R, Perne J. Approach to determine stress strain curves by FEM supported nanoindentation. Materialwissenschaft und Werkstofftechnik. 2013;**44**(6):571-576

[29] Juliano TF, VanLandingham MR, Weerasooriya T, Moy P. Extracting stress-strain and compressive yield stress information from spherical indentation. Army Research Lab Aberdeen Proving Ground MD Weapons and Materials Research; 2007

[30] Weaver JS, Kalidindi SR. Mechanical characterization of Ti-6Al-4V titanium alloy at multiple length scales using spherical indentation stress-strain measurements. Materials and Design. 2016;**111**:463-472

[31] Weaver JS, Sun C, Wang Y, Kalidindi SR, Doerner RP, Mara NA, et al. Quantifying the mechanical effects of He, W and He+ W ion irradiation on

tungsten with spherical nanoindentation. Journal of Materials Science. 2018;**53**(7):5296-5316

[32] Xiao X, Terentyev D, Yu L. Model for the spherical indentation stress-strain relationships of ion-irradiated materials. Journal of the Mechanics and Physics of Solids. 2019;**132**:103694

[33] Jin K, Xia Y, Crespillo M, Xue H, Zhang Y, Gao Y, et al. Quantifying early stage irradiation damage from nanoindentation pop-in tests. Scripta Materialia. 2018;**157**:49-53

[34] Xiao X, Chen L, Yu L, Duan H. Modelling nano-indentation of ion-irradiated FCC single crystals by strain-gradient crystal plasticity theory. International Journal of Plasticity. 2019; **116**:216-231

[35] Xiao X, Yu L. Cross-sectional nano-indentation of ion-irradiated steels: Finite element simulations based on the strain-gradient crystal plasticity theory. International Journal of Engineering Science. 2019;**143**:56-72

[36] Naveen Kumar N, Tewari R, Mukherjee P, Gayathri N, Durgaprasad P, Taki G, et al. Evaluation of argon ion irradiation hardening of ferritic/martensitic steel-T91 using nanoindentation, X-ray diffraction and TEM techniques. Radiation Effects and Defects in Solids. 2017;**172**(7–8):678-694

[37] Khosravani A, Morsdorf L, Tasan CC, Kalidindi SR. Multiresolution mechanical characterization of hierarchical materials: Spherical nanoindentation on martensitic Fe-Ni-C steels. Acta Materialia. 2018; **153**:257-269

[38] Massar C, Tsaknopoulos K, Sousa BC, Grubbs J, Cote DL. Heat treatment of recycled battlefield stainless-steel scrap for cold spray applications. Journal of Metals. 2020; **72**(9):3080-3089

[39] Pham TH, Kim SE. Nanoindentation for investigation of microstructural compositions in SM490 steel weld zone. Journal of Constructional Steel Research. 2015;**110**:40-47

[40] Yang M, Sousa B, Smith R, Sabarou H, Cote D, Zhong Y, et al. Bainite Percentage Determination and Effect of Bainite Percentage on Mechanical Properties in Austempered AISI 5160 Steel. Materials Performance and Characterization. 2021;**10**(1):110-125. DOI: 10.1520/MPC20200068

[41] Lu X, Ma Y, Zamanzade M, Deng Y, Wang D, Bleck W, et al. Insight into hydrogen effect on a duplex medium-Mn steel revealed by in-situ nanoindentation test. International Journal of Hydrogen Energy. 2019; **44**(36):20545-20551

[42] Cheng G, Zhang F, Ruimi A, Field D, Sun X. Quantifying the effects of tempering on individual phase properties of DP980 steel with nanoindentation. Materials Science and Engineering A. 2016;**667**: 240-249

[43] He B, Liang Z, Huang M. Nanoindentation investigation on the initiation of yield point phenomenon in a medium Mn steel. Scripta Materialia. 2018;**150**:134-138

[44] Nguyen NV, Pham TH, Kim SE. Characterization of strain rate effects on the plastic properties of structural steel using nanoindentation. Construction and Building Materials. 2018;**163**: 305-314

[45] Karam-Abian M, Zarei-Hanzaki A, Shafieizad A, Zinsaz-Borujerdi A, Ghodrat S. Predicting flow curves of Q&P Steel Using Sharp Pyramidal Nanoindentation on constituent phases: Isostrain method. In: Advanced Materials Research. Switzerland: Trans Tech Publ; 2021. pp. 83-91

[46] Ruiz-Moreno A, Hähner P, Fumagalli F, Haiblikova V, Conte M, Randall N. Stress- strain curves and derived mechanical parameters of P91 steel from spherical nanoindentation at a range of temperatures. Materials and Design. 2020;**194**:108950

[47] Gigax J, Torrez A, Mcculloch Q, Li N. Comparison of mechanical testing across micro- and meso- length scales [Internet]. 2019 Sep [cited 2021 Nov 29] p. LA-UR–19-29326, 1565831. Report No.: LA-UR–19-29326, 1565831. Available from: https://www.osti.gov/servlets/purl/1565831/

[48] US Patent for Instrumented indentation apparatus having indenter punch with flat end surface and instrumented indentation method using the same Patent (Patent # 10,288,540 issued May 14, 2019) - Justia Patents Search [Internet]. [cited 2021 Nov 29]. Available from: https://patents.justia.com/patent/10288540

[49] Oliver WC, Pharr GM. An improved technique for determining hardness and elastic modulus using load and displacement sensing indentation experiments. Journal of Materials Research. Jun 1992;7(6): 1564-1583

[50] Oliver WC, Pethica JB. Method for continuous determination of the elastic stiffness of contact between two bodies [Internet]. US4848141A, 1989 [cited 2021 Nov 23]. Available from: https://patents.google.com/patent/US4848141A/en

[51] Sneddon IN. The relation between load and penetration in the axisymmetric boussinesq problem for a punch of arbitrary profile. International Journal of Engineering Science. 1965; **3**(1):47-57

[52] On the generality of the relationship among contact stiffness, contact area, and elastic modulus during indentation | Journal of Materials Research | Cambridge Core [Internet]. [cited 2021 Nov 29]. Available from: https://www.cambridge.org/core/journals/journal-of-materials-research/article/abs/on-the-generality-of-the-relationship-among-contact-stiffness-contact-area-and-elastic-modulus-during-indentation/001EA7ED417E4CAD2B3C8265EC47588B

[53] Leitner A, Maier-Kiener V, Kiener D. Essential refinements of spherical nanoindentation protocols for the reliable determination of mechanical flow curves. Materials and Design. 2018; **146**:69-80

[54] Tabor D. The Hardness of Metals. Oxford, New York: Oxford University Press; 2000. p. 192 (Oxford Classic Texts in the Physical Sciences)

[55] Cheng YT, Cheng CM. Scaling, dimensional analysis, and indentation measurements. Materials Science & Engineering R: Reports. 2004;**44**(4): 91-149

[56] Koeppel BJ, Subhash G. Characteristics of residual plastic zone under static and dynamic Vickers indentations. Wear. 1999;**224**(1):56-67

[57] KLA_AppNote_Stress-Strain_Ti.pdf [Internet]. [cited 2021 Nov 29]. Available from: https://www.kla-tencor.com/wp-content/uploads/KLA_AppNote_Stress-Strain_Ti.pdf

[58] Gleason MA, Sousa BC, Tsaknopoulos K, Grubbs JA, Hay J, Nardi A, et al. Application of mass finishing for surface modification of copper cold sprayed material consolidations. Materials. 2022;**15**(6): 2054

www.ingramcontent.com/pod-product-compliance
Lightning Source LLC
Chambersburg PA
CBHW081541190326
41458CB00015B/5611
9781839699603